Hidden in Plain Sight 12

Andrew Thomas studied physics in the James Clerk Maxwell Building in Edinburgh University, and received his doctorate from Swansea University in 1992.

His *Hidden In Plain Sight* series of books are science bestsellers.

Also by Andrew Thomas:

Hidden In Plain Sight
*The simple link between relativity
and quantum mechanics*

Hidden In Plain Sight 2
The equation of the universe

Hidden In Plain Sight 3
The secret of time

Hidden In Plain Sight 4
The uncertain universe

Hidden In Plain Sight 5
Atom

Hidden In Plain Sight 6
Why three dimensions?

Hidden In Plain Sight 7
The fine-tuned universe

Hidden In Plain Sight 8
How to make an atomic bomb

Hidden In Plain Sight 9
The physics of consciousness

Hidden In Plain Sight 10
How to program a quantum computer

Hidden In Plain Sight 11
The logic of consciousness

HIDDEN

IN PLAIN SIGHT 12

Consciousness and the Steam Engine

ANDREW THOMAS

Hidden In Plain Sight 12

Copyright © 2021 Andrew D.H. Thomas
hiddeninplainsightbook@gmail.com

All rights reserved.

ISBN: 9798597795256

CONTENTS

PREFACE

In the *Hidden in Plain Sight* series of books, this is my third book dedicated to the subject of consciousness.

In my first book on the subject, it was explained how information is likely to play a central role in the emergence of consciousness. Then, in my second book on the subject, it was suggested that the physics of thermodynamics might hold the key to unlocking the problem. This third book on the subject of consciousness builds on the conclusions of those previous two books (information and thermodynamics).

This book represents the culmination of my three years working on this problem. I am pleased to say that in this third book I now feel confident to present an original, scientific theory of consciousness. The theory is being presented in this book for the first time.

As you will see, the theory is very simple. It attempts to describe the basic principles of consciousness, explaining why some devices are capable of generating consciousness, whereas other devices do not possess that special quality. Unusually for this subject, the theory is firmly based in mainstream, orthodox physics.

I believe the theory is correct.

Of course, I am totally biased in favour of my own theory! But by the time you finish this book, I hope you will agree with me.

Andrew Thomas (hiddeninplainsightbook@gmail.com)
Swansea, UK,
2021.

*"All our actions,
from digestion to artistic creation,
are at heart captured by the essence
of the operation of a steam engine."*

\- PROF. PETER ATKINS
"Four Laws that Drive the Universe"

1

THE INDUSTRIAL REVOLUTION

The eighty-year period from 1760 to 1840 was a time of great change in Britain. It was a period of rapid technological innovation, and the emergence – for the first time in human history – of a modern industrial economy. In this chapter we will be examining how a small, rain-soaked island in the North Atlantic became the world's dominant industrial power, and the richest nation on Earth. It is the story of the Industrial Revolution.

The Industrial Revolution was based on a series of inventions. The inventors tended to be based in the North of England rather than the more genteel South. These men were not scientists: hardly any of them had received a university education. They had no interest in knowledge for knowledge's sake. No, they were practical men, often with an engineering background, who could recognise problems which urgently required solutions. Usually, their motivation for inventing was purely financial.

Roger Osborne considered the impact of these inventors in his book about the Industrial Revolution, *Iron, Steam &* *Money*:

The Industrial Revolution came about because of inventors and their innovations. So, having survived a few decades in the wings of history, the great inventors – Watt, Newcomen, Hargreaves, Arkwright, Trevithick – are once again centre stage as the great movers in this historical drama. It was a single generation of British artisans who made possible Britain's transition to industrialisation and transformed the prospects of humanity.

Let us start by considering the invention which Roger Osborne called "one of the greatest inventions in human history". It is also the invention which – as we shall see later in this book – potentially holds the key to unlocking the mystery of consciousness.

It is the invention of the steam engine.

James Watt and the cooking pot

Every British schoolchild knows the legend of how James Watt invented the steam engine. Indeed, it was the story I was taught in my history lessons in school. If you do not already know the story, allow me to tell you the legend of James Watt and the cooking pot.

The various forms of the legend revolve around a young James Watt watching a cooking pot which belonged to his mother. As the water in the pot boiled, it turned into steam. The only way the water could escape from the pot was by lifting the metal lid which was on top of the pot. As we can imagine, the escaping steam caused the lid of the pot to rapidly jump up and down. According to legend, it was at this moment that James realised that there was "power in steam".

The following drawing shows an older James Watt watching the moving lid of the legendary cooking pot (from an original engraving by Jacob Abbott, 1854):

We will now examine the "jumping" of the cooking pot lid in more detail. In doing so, we will discover how it is possible to convert the intermittent lifting of the cooking pot lid into a continuous form of motion. The method will involve a linked sequence of steps, a sequence which lies at the heart of all steam engines.

The sequence is known as a *cycle*.

The cycle

Let us consider the jumping of the cooking pot lid in more detail. The three steps are shown in the following diagram:

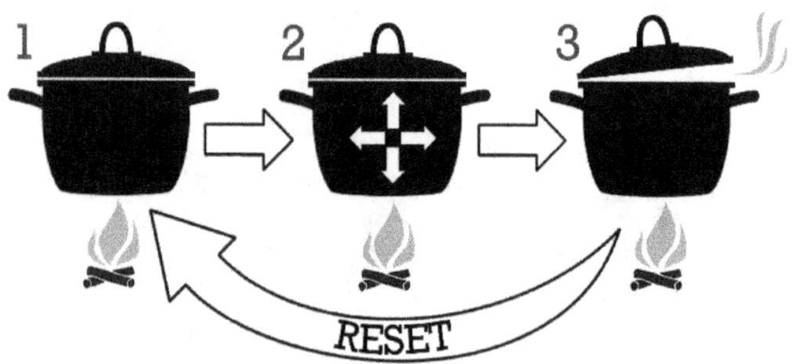

Step One of the previous diagram shows the cooking pot above the fire, with the lid lying on top of the pot. At this stage, the pressure of the steam in the pot is the same as the atmospheric pressure outside the pot.

Step Two shows pressure building inside the pot as the water is heated, creating steam.

Step Three shows the cooking pot lid flying up off the pot. This is caused by the pressure of the steam inside the pot becoming sufficiently large to lift the lid. However, once the lid lifts off the pot, the pressure inside the pot is equalised with the atmospheric pressure outside the pot, so the lid drops down again. The system is therefore returned to its initial state.

This last step is vital. It acts as a "reset" operation (drawn as a curved arrow on the diagram). The result of the reset operation is to perfectly restore the state of the system back to its initial state (in this case, equalising the pressure inside the pot with the atmospheric pressure). Because of the reset operation, and the restoration of the system back to its initial state, it is possible for the entire sequence of steps to immediately repeat.

The sequence of steps then returns to Step One again, and the sequence repeats with pressure building inside the pot again. This sequence of steps therefore forms a cycle. A cycle is defined as being a sequence of steps performed by a system before the system returns to its initial state. The cycle then repeats continuously, creating continuous motion.

A repeated cycle plays a central role in the operation of a steam engine, with the motion of a piston moving in and out of a cylinder representing one cycle. As we shall see, one particular cycle will play an important role towards the end of this book when we come to consider human consciousness.

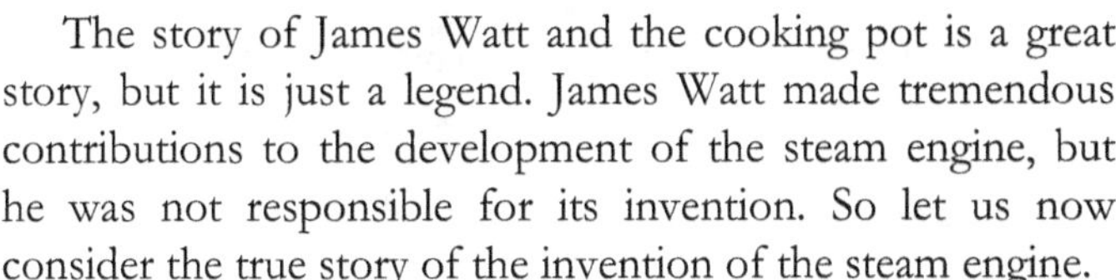

The story of James Watt and the cooking pot is a great story, but it is just a legend. James Watt made tremendous contributions to the development of the steam engine, but he was not responsible for its invention. So let us now consider the true story of the invention of the steam engine.

The first steam engine

Britain in the 1700s found itself in the midst of an energy crisis, a crisis which threatened to stop the Industrial Revolution before it had even begun. The problem was that, up to that point, vast forests had provided a plentiful supply of wood which could be burnt as fuel. With the expanding population, however, the demand for wood had increased to such an extent that Britain had become largely deforested. The country was rich in coal, which could potentially be used as an alternative source of energy. However, to extract coal from deeper underground involved digging mines which rapidly filled with water. Some method had to be found for pumping the water out of deep mines.

The solution was found by Thomas Newcomen. Newcomen lived in Cornwall, which is a mineral-rich county in the far south west of England. It was Newcomen who invented the world's first steam engine, and it was used to pump water out of mines in Cornwall.

In the previous section, it was considered how pressurised steam might be used to power a steam engine. The idea seems intuitively obvious, so it is surprising that Newcomen's first steam engine was powered by a completely different principle. In fact, it might be said that Newcomen's engine worked by the reverse principle, exploiting the power of low pressure (a partial vacuum) rather than using pressurised steam.

The three stages in the cycle of a Newcomen steam engine are shown in the following diagram:

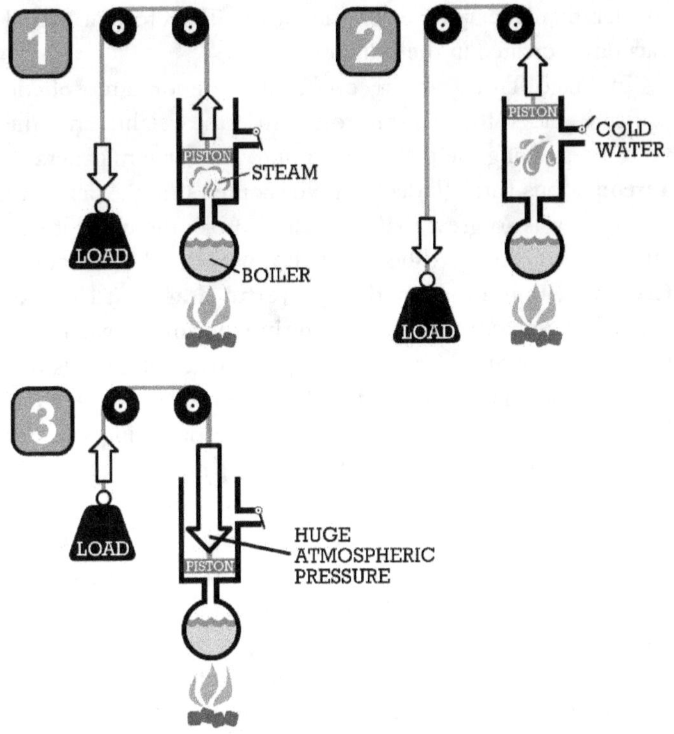

The previous diagram shows a round boiler containing water being heated by burning coal. In Stage One of the cycle, it can be seen that the load is allowed to drop, and this lifts a piston which is in a cylinder. As the piston rises, it draws steam from the boiler into the cylinder. Note that the steam engine is not performing work at this stage of the cycle as the load is being allowed to drop (i.e., the engine is not lifting the load).

In Stage Two of the cycle, it can be seen that cold water is injected into the cylinder. This causes the steam to condense into water (just as your breath condenses into water on a cold pane of glass). When steam condenses and turns into water, the volume of water created is 1,800 times smaller than the volume of the steam. Therefore, a partial vacuum is created in the cylinder.

In Stage Three of the cycle, the exterior atmospheric pressure plays the central role. Air has weight, and the pressure pushing the piston down into the vacuum generates a tremendous force (indeed, in Newcomen's first experiment the force was so great that the piston broke the chain it was attached to and crushed the bottom of the cylinder). **Crucially, it is only on this powerful downward stroke that the Newcomen steam engine performs work.** As you can see in Stage Three of the previous diagram, it is this powerful downward stroke that lifts the heavy load (such as lifting a load of coal out of a mine, or operating a water pump to raise water from the bottom of a deep mine).

Steam engines such as this which only perform work on either the up or down stroke – but not both – are called *single-acting* engines. The internal combustion engine in your car or motorcycle is another example of a single-acting engine, with a piston only performing work on the down-stroke when the spark plug above the piston fires.

The three-stage cycle of the Newcomen engine then repeats. However, unlike James Watt's jumping cooking pot lid, this was most certainly not a rapid repeat: a Newcomen steam engine could only perform about twelve strokes per minute.

So that is how the world's first steam engine operated. Roger Osborne describes Thomas Newcomen's achievement in his book *Iron, Steam & Money*: "It deserves a place among the greatest inventions in human history. This was the first machine that replaced human, animal, and water power; its effect on the world could hardly have been greater."

The fiery monster

Before the invention of the steam engine, the power for factories was mainly produced by waterwheels. One advantage of a steam engine as compared to a waterwheel is that a steam engine is mobile. It was realised that it might be possible to take advantage of the mobility of a steam engine to put a steam engine on wheels – thereby creating a steam locomotive. A big problem, however, was the huge size and weight of the Newcomen engine. If steam engines were to be used for locomotion, they had to be made smaller and lighter.

As explained in the previous section, the Newcomen engine used a vacuum and atmospheric pressure to produce power. Because of its use of a vacuum, it might be called a low pressure engine. However, it was realised that steam at high pressure could also be used to move a piston (in the same way that steam moved the lid of James Watt's legendary cooking pot).

James Watt was, by now, a successful inventor who had made significant improvements to Newcomen's engine. However, Watt was reluctant to make the change to using high pressure steam because of safety concerns: there had been several cases in which the boilers of high-pressure engines had exploded, causing several fatalities.

James Watt's assistant was an enterprising young engineer named William Murdoch. Ignoring Watt's concerns about its safety, Murdoch became interested in the possibilities of high pressure steam. To test its feasibility, Murdoch built a small working model of a steam locomotive. The resultant model had three wheels, and was about a foot in height. The model has survived to this day, and is shown in the following photograph:

Murdoch was living in Redruth in Cornwall at the time, and he found the perfect track on which to test his vehicle. The path which led from the church had steep banks on either side, which would act to keep his vehicle on track. Murdoch started the engine, and ran alongside it as it powered its way towards the church. In his book *Energy and the Unexpected*, Keith Laidler described what happened next:

> *When he made his first trial, at night, the engine moved so fast that it out-distanced him and almost ran down the rector of the church who happened to be walking along the road. As he ran away in terror, the rector assumed the fiery monster to be the embodiment of the devil.*

All initial experiments with steam-powered locomotion proved unsuccessful due to the poor quality of the roads, and William Murdoch did not pursue the idea. However, Murdoch had a neighbour in Redruth named Richard Trevithick who was also an engineer. Murdoch showed his working model to Trevithick who was intrigued, and Trevithick decided to build his own steam locomotive.

Trevithick realised that a solution to the problem of rough road surfaces would be to put the locomotive on the metal rails which were being used in the coal and iron industry. As a result, on 21st February 1804, Trevithick demonstrated the first steam locomotive running on the rails of the Penydarren Ironworks in Merthyr Tydfil in South Wales. It travelled a distance of 9.75 miles carrying ten tons of iron and seventy people, travelling at an average speed of 2.4 miles per hour. As to why the average speed was so slow, Trevithick himself described the journey:

> *We carry'd ten tons of iron, five waggons, and 70 men riding on them the whole of the journey. It's above nine miles which we perform'd in 4 hours and 5 mins, but we had to cut down some trees and remove some large rocks out of the road. The engine, while working, went nearly 5 miles per hour.*

At this point, I must declare a personal interest as Merthyr Tydfil is the town in which I grew up. It is a remarkable town, with a rich industrial heritage. The achievement of Richard Trevithick has not been forgotten and a monument was erected in the town centre. A model of Trevithick's locomotive sits proudly atop the monument:

The inscription on the monument reads:

Richard Trevithick, 1771-1833. Pioneer of high pressure steam built the first steam locomotive to run on rails. On February 21ˢᵗ 1804 it traversed the spot on which this monument stands on its way to Abercynon.

I no longer live in Merthyr, having moved to the town of Swansea which is thirty miles from Merthyr. However, in the National Waterfront Museum in Swansea, a full-size completely operational replica of Trevithick's locomotive has been built. In the following video taken by Matt Slade, you can see the locomotive in action, running on the rails outside the museum in Swansea:

http://tinyurl.com/steamtrainvideo

That first journey of Trevithick's locomotive was a truly monumental event, not just for Britain but for the world. According to Roger Osborne in *Iron, Steam & Money*: "The Penydarren locomotive had most of the essential features of steam locomotives that would transform the world over the next 150 years."

From a physics viewpoint, Trevithick's locomotive showed it was possible to convert heat into motion, the heat in the boiler being converted into the motion of the locomotive. Therefore, in the 1800s, what was needed was a new science which was able to describe the connection between heat and motion. That science was the science of *thermodynamics*. It is that new science of thermodynamics which we shall consider in the next chapter.

2

THE MOTIVE POWER OF FIRE

Our attention now turns across the English Channel to Britain's neighbour, France.

During the period of the Industrial Revolution, Britain was involved in a series of lengthy wars with France. The leader of France at that time was the ambitious and charismatic Napoleon Bonaparte. Napoleon was a brilliant military commander who developed novel military strategies which laid the foundations of modern warfare. Instead of the traditional head-on approach to battle, Napoleon's armies were fast and mobile, capable of outflanking the opposing forces.

Napoleon was famously a man of short stature, and it has been suggested that he overcompensated for his lack of height by presenting himself as a figure of great grandeur (in psychology, this has even been given the term the *Napoleon complex*). Indeed, by 1804, Napoleon had awarded himself the title of first Emperor of France. The following painting by Jacques-Louis David titled *Napoleon Crossing the Alps* shows Napoleon in a typically grandiose pose (Napoleon actually crossed the Alps seated on a mule):

Napoleon enjoyed astounding military success in a series of famous battles, most notably the defeat of the much larger Russian and Austrian armies at the Battle of Austerlitz in 1805. At its height, the French Empire under Napoleon extended from Portugal to Moscow.

Napoleon's progress across Europe appeared unstoppable. However, on the 18th June 1815 in the fields of Belgium, the forces of Napoleon met the forces of his greatest enemy, the British army led by the Duke of Wellington. The battle is now known by the name of a nearby village: Waterloo.

The night before the battle saw a thunderstorm of biblical proportions. The torrential rain made the ground soft and boggy. As a result, the French forces were unable to get their cannons into position until late in the morning. Also, because of the soft ground, the recoil from the cannons caused them bury themselves into the ground after firing. A huge effort then had to be made to haul the cannons back into position, which greatly reduced the rate of firing. Unable to rely on his artillery to devastate the British forces from a distance, Napoleon's forces had to engage the British infantry in close quarters fighting. After a brutal day of hand-to-hand fighting, Napoleon's forces were narrowly defeated by the British army.

The Battle of Waterloo was to be Napoleon's final defeat. He was deposed as Emperor, and exiled to the small, damp island of St. Helena in the South Atlantic – one of the most remote islands in the world – where he could pose no more threat. Napoleon died on St. Helena six years later, supposedly from boredom.

The humiliation of Sadi Carnot

In 1800, soon after rising to power in France, Napoleon appointed Lazare Carnot to be his Minister of War. Carnot was trained as a military engineer, and was known for his analytical mind. Throughout his leadership of France, Napoleon came to rely on the strategic advice of Carnot. It was Carnot who developed the modern principles of lightning pace combined with an outflanking strategy (*pincer movement*) which Napoleon employed with great success.

Perhaps we should not be surprised that Carnot had such a strategic and technical mind as he was also a highly-talented physicist. When he was not involved in military activities, Carnot continued to develop his ideas about physics, and he published several books about physics in his spare time.

Lazare Carnot had a son whose name was Sadi. Carnot had often taken the young Sadi to meet Napoleon. It is recorded that a four-year-old Sadi once ran up to Napoleon shaking his fist and reprimanding the continent-conquering

Emperor for his rude treatment of women. Luckily, Napoleon roared with laughter.

When the time came to choose a college for Sadi, with his family background in physics and the military, there was an obvious choice. An establishment had been created precisely for the purpose of fostering the scientific talents of young French men. The French had realised that British military superiority had been founded on their technological superiority gained during the Industrial Revolution. It was therefore decided that French engineers and scientists needed the best training. As a result, in 1794, Lazare Carnot co-founded the École Polytechnique in the Latin Quarter of Paris as the premier training school for young scientists.

Here is a painting of Sadi Carnot in 1813 at the age of seventeen, wearing the traditional uniform of a student at the École Polytechnique (painting by Léopold Boilly):

The École Polytechnique became the first great scientific school of the modern world. The roll call of those associated with the school now reads like a list of the early giants of mathematical physics, including Fourier, Lagrange, Laplace, and Ampère. The terms "Fourier analysis", "Lagrangian", and "Laplace transform" are now well known to every physics student.

The name of Sadi Carnot was added to that illustrious list of names when he enrolled at the École Polytechnique in 1812 at the age of sixteen. However, by the time Sadi had finished his studies in 1814, Napoleon was in full retreat and Paris was besieged. The students of the college – including Sadi Carnot – fought bravely to defend the outskirts of the city, but were forced to retreat under sustained artillery fire.

The final defeat of France by the British at Waterloo in 1815 had a devastating impact on Sadi Carnot. Remember, Sadi Carnot's father had been Napoleon's Minister of War, and he had become one of Napoleon's finest generals. Sadi Carnot became determined to avenge the defeat of his father – and his country.

The revenge of Sadi Carnot

When Sadi Carnot saw a steam engine for the first time, he became obsessed with the question of understanding how it worked. The problem was that the operation of a steam engine was not fully understood. Isaac Newton had presented his laws of mechanics a century earlier, but Newton had no understanding of energy, or heat. With no theory to guide them, steam engine designers had proceeded on the basis of trial-and-error: build a different engine design, and see if it produced more power. A theory of steam engines was therefore desperately needed.

In 2012, the BBC made a two-part documentary series about thermodynamics called *Order and Disorder*. The series was written and presented by Prof. Jim Al-Khalili. Here is a link to a video of episode one on YouTube:

http://tinyurl.com/videoheat

The video is excellent and is well worth watching.

At the 12-minute mark of the video, Jim Al-Khalili considers Sadi Carnot's reaction to the French military defeat:

> *The humiliation he felt personally would drive him and motivate him to uncover a profound insight into how all engines work.*
>
> *Carnot came from a highly-respected military family. After the French defeat, he became determined to reclaim French pride.*

Jim Al-Khalili then proceeds to describe how Sadi Carnot realised the true basis for the British military advantage:

> *What really bothered Carnot was the technological superiority that France's enemies seemed to possess. And Britain, in particular, had this huge advantage both militarily and economically because of its mastery of steam power. So Carnot vowed to try and understand how steam engines work, and use that knowledge for the benefit of France.*

In the video, Prof. Simon Schaffer then describes how Sadi Carnot realised a method to gain revenge over Britain:

> *He says absolutely explicitly that if you could take away steam engines from Britain then the British Empire would collapse.*

Sadi Carnot became determined to exact revenge on Britain, and the way he would do it would be to design a better steam engine … for France.

Historic theories of heat

The problem faced by Sadi Carnot was that the operation of a steam engine was based on heat, and nobody actually knew what heat was. The ancient Greeks believed that everything was composed of four fundamental elements: earth, air, fire, and water. Fire was thus thought of as a substance.

This model of the elements survived through the Middle Ages and into the Renaissance. However, in 1678, the German alchemist Johann Becher proposed a new theory called the *phlogiston* theory. According to Becher, all inflammable substances contained the material known as phlogiston. When the material burnt, the phlogiston was released, which gave the appearance of a flame. The material which was left (the ash) was then consider to be "dephlogisticated". Hence, the ash was considered to be the true essence of the material. Growing plants then absorbed the phlogiston from the air, which explained why they burned so well.

The phlogiston theory was generally accepted for 100 years, but flaws in the theory started to emerge. Firstly, the theory predicted that materials should lose weight as they burned and the phlogiston was released. However, some metals actually increased in weight when they burned (we now understand that this is due to the metal combining with oxygen from the air). Secondly, the theory had no explanation of why heat should flow in only one direction: from a hot object to a cold object. An improved theory was required.

As a result, in 1783 the great French chemist Antoine Lavoisier introduced the *caloric* theory of heat. The theory proposed that heat consisted of a weightless substance called caloric. According to the theory, caloric was a material substance (a fluid) and the amount of caloric in the universe was constant: caloric could be neither created nor destroyed. It was also proposed that caloric was self-repellent. The self-repellent nature of caloric provided an explanation as to why heat travels only from hot objects to cold objects: caloric would naturally move from regions in which it was highly concentrated (hot) to regions in which it was more sparsely concentrated (cold).

Although we now know the caloric theory to be incorrect, it gave Sadi Carnot the idea he needed to be able to analyse steam engines, as we shall now see ...

The waterwheel of heat

Sadi Carnot wondered if there was a limit to the amount of power which could be produced by a steam engine. Assuming that there was a limit to the efficiency of a steam engine, then what was the formula for that efficiency?

Carnot's analysis of steam engine efficiency was heavily influenced by waterwheels. Before the widespread adoption of the steam engine, factories had generated their power from waterwheels. Waterwheels provided a constant, smooth rotary action, ideal for driving the mill machinery used for spinning cotton in the Industrial Revolution. The use of waterwheels might sound primitive, but we still use waterwheels today – though we call them hydroelectric plants. Hydroelectricity currently provides sixteen percent of the world's electricity, and the world's largest power plants are hydroelectric plants.

Let us analyse how power can be produced by a waterwheel. The following diagram shows how water flows from a height, h, with the flow of water powering the wheel:

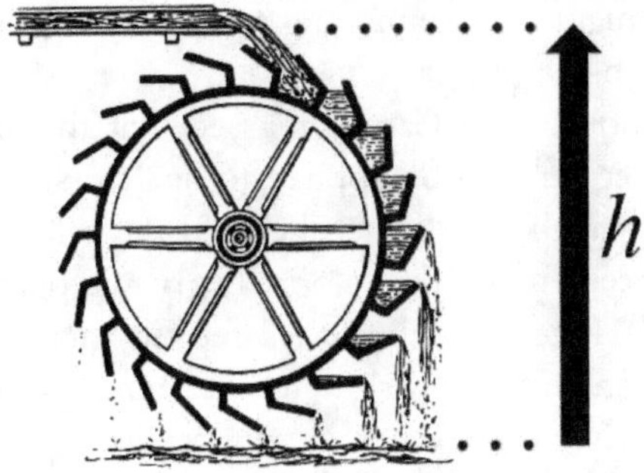

You may remember from physics lessons in school that an object which is at a height contains *potential energy* due to that height. That means that the water at the top of the waterwheel will have more energy than the water when it drops to the bottom of the waterwheel. But, according to the law of conservation of energy, energy cannot just vanish – it must have gone somewhere. So if the energy of the water is less at the bottom of the waterwheel than it was at the top, then where has the energy gone?

To answer that question, we need to consider what we mean by "work". "Work" has a very specific meaning in physics. In physics, work means the application of a force to move an object. So, for instance, if a pump lifts water out of a deep mine, then that pump has performed work.

Now let us consider the definition of energy. What is energy? Energy is defined as the ability of a system to do work. If a system has more energy, it can do more work. Exactly the same principle applies to human beings: if you are feeling more energetic, you can do more work. If you have more energy, you can do the gardening, or lift some heavy rocks.

So, as the energy of the water is less at the bottom of the waterwheel than it was at the top, we can now understand where the missing energy has gone. The missing energy from the waterwheel is used to perform work inside the factory.

That work might involve moving machinery such as a cotton weaving loom, for example.

At this point, Sadi Carnot realised that this model of a waterwheel could also be applied to analyse steam engines. The breakthrough came when he used the caloric model of heat. If you remember, the caloric theory said that heat was a substance (a fluid), which always moved in one direction: from high temperature to low temperature. Carnot realised that the behaviour of water in a waterwheel was identical: water was a substance which always moved from a high level to a low level. Therefore, the passage of heat through a steam engine was the same in principle as the passage of water through a waterwheel.

Carnot realised he could describe any steam engine in abstract form as a form of "waterwheel" with the caloric fluid of heat "falling" from a high temperature to a cold temperature. The high temperature would be the temperature of the hot boiler, and the low temperature would be the temperature of the external environment.

How, then, could Carnot use this model to analyse the efficiency of a particular steam engine? Well, as has just been described, the potential energy of a waterwheel is dependent on the height difference of the water in the two reservoirs. Carnot applied similar logic to his analysis of the steam engine. In the case of the steam engine, the available energy would be dependent on the difference in temperatures.

In that case, Carnot realised that the available energy which could be produced by the steam engine would depend on the difference in temperatures:

$$T_H - T_L$$

where T_H is the high input temperature, and T_L is the lower output temperature. This formula might be considered as representing the height of the "waterwheel of heat".

Does this value then represent the efficiency of the steam engine? Not quite. To get to the final form of the formula, let us consider what we mean by "efficiency". Efficiency is the proportion of the energy which is input to the system which is converted into work. We have just seen that the energy which is converted into work is proportional to T_H–T_L, so to get the efficiency we have to divide that value by the amount of energy which was input to the system which would be the total height of the waterfall, T_H. The final formula for the efficiency of a steam engine is then:

$$\frac{T_H - T_L}{T_H}$$

Even though Carnot was using the caloric model of heat – which we now know is incorrect – it is amazing that this formula remains the correct formula for the efficiency of a steam engine.

In 1824, Sadi Carnot published his results about the efficiency of steam engines in a book titled *Reflections on the Motive Power of Fire*. This classic book is now considered to have created the science of thermodynamics, the science of heat ("thermo") and motion ("dynamics").

Here is the cover of Sadi Carnot's legendary book. You can see that Carnot described himself as an "Ancien élève de l'École Polytechnique", which translates to "Former pupil of the École Polytechnique":

RÉFLEXIONS

SUR LA

PUISSANCE MOTRICE

DU FEU

ET

SUR LES MACHINES

PROPRES A DÉVELOPPER CETTE PUISSANCE.

PAR S. CARNOT,

ANCIEN ÉLÈVE DE L'ÉCOLE POLYTECHNIQUE.

A PARIS,

CHEZ BACHELIER, LIBRAIRE,

QUAI DES AUGUSTINS, N°. 55.

1824

Heat engines

What does Carnot's formula tell us about how to improve the efficiency of a steam engine? First, allow me to remind you of Carnot's formula:

$$\frac{T_H - T_L}{T_H}$$

Now let us put some numbers into the formula. Let us imagine a steam engine which has a boiler full of boiling water (which is 100 degrees Celsius, equal to 373 degrees Kelvin). Let us also imagine that the external temperature is freezing (which is 0 degrees Celsius, equal to 273 degrees Kelvin). Putting those numbers into the formula gives a value for the efficiency of that steam engine:

$$\frac{373 - 273}{373} = 0.27$$

So the efficiency of that steam engine is 0.27, or 27%. If you think that sounds inefficient, well, you would be correct. But, considering the formula, how could the efficiency be improved? Before Carnot's result, it had been thought that engine efficiency could be improved by using different materials, such as boiling alcohol instead of water. However, Carnot's formula revealed that the important factor was the working temperature – not the materials used. In order to improve the efficiency of this steam engine, we need to increase its working temperature, T_H.

Prof. Jim Al-Khalili considers this strategy at the 17-minute mark of the excellent BBC thermodynamics video (a link was provided earlier in this chapter). According to Jim Al-Khalili:

Carnot's crucial insight was to show that to make any heat engine more efficient, all you had to do was to increase the difference in temperature between the heat source and the cooler surroundings.

Ultimately, a car engine is more efficient than a steam engine because it runs at a much hotter temperature.

Note the use of the term "heat engine" in the previous quote by Jim Al-Khalili. The term "heat engine" is the generic term for any engine which exploits the passage of heat from hot to cold, with the difference in heat energy being used to perform mechanical work:

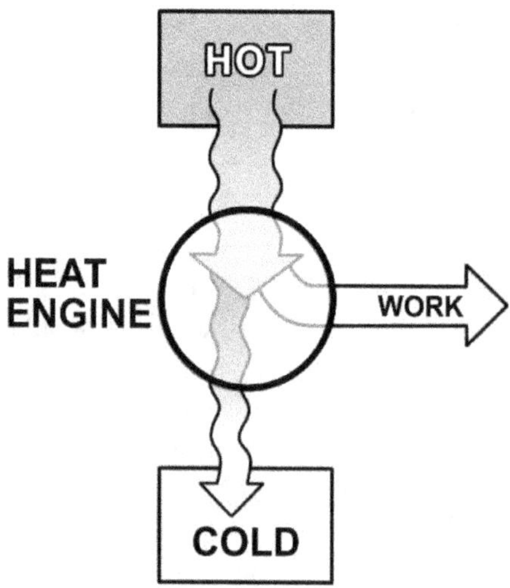

A heat engine is powered by the movement of heat from hot to cold. For example, in a steam engine that would be the movement of heat from a hot boiler, through the steam engine, and then expelled as exhaust steam.

A steam engine is just one example of a heat engine. Many other types of heat engine exist, all working on the same principle. For this book, I have created an entertaining video showing some other types of heat engine (you may be surprised!).

Here is a link to the video:

http://tinyurl.com/hotpower

I hope you enjoy the video!

3

REVERSIBILITY

In his 1687 masterwork, the *Principia*, Isaac Newton had presented his laws of mechanics. These were the laws of masses, forces, and accelerations. Newtonian mechanics provided the theoretical tools needed by the engineers who created the Industrial Revolution. Just as importantly, Newton's discoveries generated a new excitement in the general population about the possibilities of science, and this encouraged engineers and entrepreneurs such as James Watt and Richard Trevithick to devise their world-changing inventions. It was largely because of Isaac Newton that Britain became the world's leading technological nation.

As a result, Newton became an almost godlike figure in Britain. The insights of Newton, which had been captured in the *Principia*, were treated with almost the same reverence as the Bible.

When Newton died, the English poet Alexander Pope wrote the following famous epitaph which reflected the almost divine status of Newton:

> *Nature, and nature's laws lay hid in night;*
> *God said "Let Newton be" and all was light.*

The esteem in which Newton was held is clearly displayed in his magnificent tomb at Westminster Abbey:

On the tomb, you can see the reclining Newton resting on four of his books, including the *Principia*. At the top of the tomb can be seen the celestial globe onto which is drawn the path of Newton's Comet of 1680.

It was Newton who had shown that the same force which holds the planets in their orbits was also responsible for pulling an apple down from a tree. Previously, these two phenomena were considered to be completely unrelated. By showing that they were two different aspects of the same force of gravity, Newton had achieved one of the first *unifications* in physics.

Beneath the sculpture of Newton on his tomb, the heliocentric model of the Solar System is drawn showing the Sun at its centre and the known planets orbiting around the

Sun – according to Newton's law of gravitation. By unlocking the mystery of the motion of the planets and the heavenly bodies, it finally seemed as though mankind had completely unlocked the secrets of the universe.

So when the steam engine appeared in the 1800s, it might have been expected that it would be no problem for Newton's laws to explain the behaviour of the steam engine. After all, the laws which described the motion of planets and comets could surely explain the operation of a lump of iron filled with water and heated by coal.

What is more, in the previous chapter we have seen how the operation of a steam engine is really quite simple. Like all heat engines, it is based of the one-way flow of heat energy, always from hot to cold, and never in the opposite direction.

But there was only one problem. The laws of Isaac Newton were not one-way at all.

The laws of Isaac Newton were completely reversible.

Reversibility in the movies

Newton's three laws of motion form the basis of Newtonian mechanics (*classical mechanics*). The three laws are as follows:

1) An object remains at rest or continues at constant velocity unless it is acted on by a force. This is sometimes called the *law of inertia*.

2) A force will cause an acceleration of an object. The acceleration will be proportional to the magnitude of the force, and inversely proportional to the mass of the object.

3) Every action (force) will produce an equal and opposite reaction.

The three laws are clearly very simple, but they govern the motion of all objects. Part of the simplicity of the laws arises because they include no mention of any particular direction of time. The laws do not specify that they only apply in the forward time direction. In other words, there is nothing in the laws to indicate that they would not apply in a system in which all the objects suddenly reversed their direction of motion and started running backwards – as if time had been reversed. It is therefore said that Newton's laws are *time-reversal invariant*.

There is one simple way to demonstrate the reversibility of Newton's laws, and that is because it is possible for us to generate an artificial form of "time reversal". And the way we can achieve that time reversal is by making a movie.

The British movie director Christopher Nolan has gained a reputation for making innovative, unconventional movies which consider deep themes about the nature of time, space, and reality. His movies often incorporate ideas from science, particularly physics. Nolan's desire for scientific accuracy was most clearly demonstrated in his movie *Interstellar* which relied on technical advice from the Nobel prize-winning theoretical physicist Kip Thorne.

Nolan first showed an interest in the phenomenon of time reversibility in his movie *Memento*. This brilliant movie told the story of a man with amnesia trying to find the murderer of his wife. But what made the movie so remarkable is that the screenplay unfolded in reverse, starting with the final scene, and ending with the first scene. In other words, the events in the movie unfolded in time-reversed order (though Nolan still manages to surprise us at the end of the movie – or should I say the start?).

But it was in a later movie that Christopher Nolan's interest in time reversal was to play a central role. That movie was *Tenet*, released in the late summer of 2020.

Unfortunately, the coronavirus pandemic, and the resultant closure of cinemas, meant that the release date of

Tenet was delayed several times. However, the movie was finally released in the UK at the end of August. And so it was that I found myself in a virtually empty cinema with just four other socially-distanced people.

The central premise of *Tenet* is that there exists a technologically-advanced device (called a "turnstile") which can reverse the direction of the flow of time for objects. This idea resulted in some spectacular and inventive time-reversed scenes. As was stated earlier in this section, time reversal would involve objects in a system reversing their direction of motion. Hence, as an example, one time-reversed scene in *Tenet* shows bullets leaving a target and following a reversed trajectory back into the barrel of a gun.

These time-reversed motions shown in movies such as *Tenet* are the best way to demonstrate the reversibility of Newton's laws. All we need to do is make a movie of the movements and interactions between a group of objects, and then play that movie backwards. If it can be seen that Newton's laws still apply to the reversed sequence of events, then that means that Newton's laws are reversible in time. As the physicist Paul Davies says in his book *The Physics of Time Asymmetry*: "Time reversal may be imagined as taking a movie film of the original motion, and then playing it backwards."

The following diagram shows three frames of a short movie. The movie is of a game of snooker (or pool). The first frame shows a stationary black ball in the middle of the frame, and a moving white ball coming in from the left of the frame. The second frame shows the white ball colliding with the black ball. The third frame shows the result of the collision: the white ball becomes stationary, while the black ball moves off at speed:

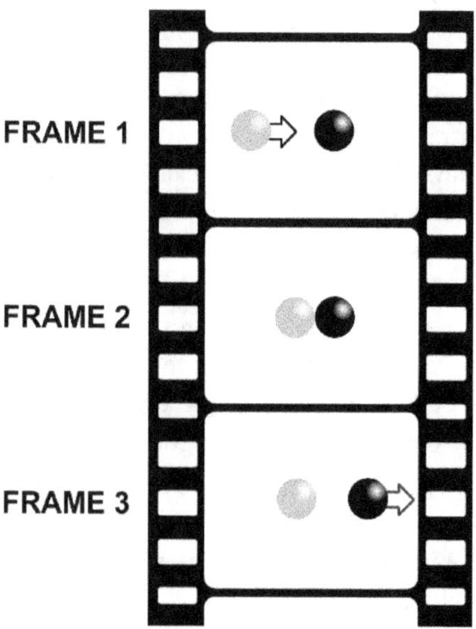

Simple motions of balls, and collisions (forces) between those balls, are described by Newton's laws of motion. So let us reverse this short movie and see if it still makes sense, i.e., do Newton's laws still apply to the time-reversed motions?

To answer that question, let us consider the sequence of frames in the reverse order, so that the last frame becomes the first frame, and the first frame becomes the last frame.

Here is the new sequence of frames, now numbered from Frame 3 to Frame 1:

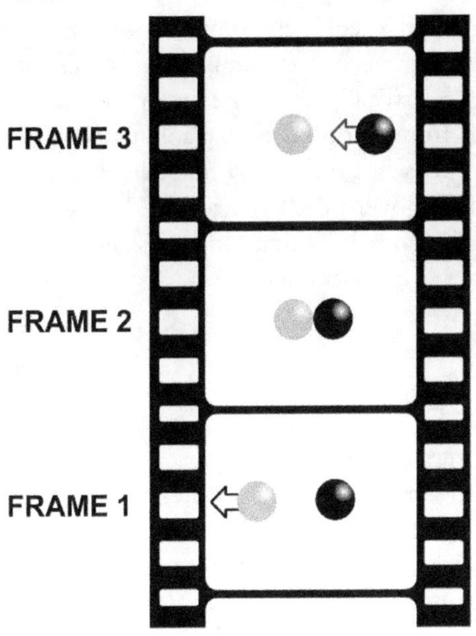

As you can see, the first frame now shows the white ball stationary in the middle of the frame, and the black ball moving in from the right of the frame. The second frame shows the collision between the two balls. And the new final frame (the old first frame) shows the black ball stationary in the middle of the frame, while the white ball moves out of the frame to the left.

So this new time-reversed sequence of events seems reasonable, perfectly in accordance with Newton's laws of motion. This tells us that Newton's laws are reversible in time.

However, the situation would have been different if our movie had involved the passage of heat, rather than the motions of snooker balls. Imagine a movie of a hot object in a cold room. As the movie is filmed, heat leaves the object, and passes into the colder environment. Now let us imagine we play that movie in reverse. In reverse, we would see heat passing from the cold environment to heat the warmer object, thereby raising its temperature. This behaviour is something we would never encounter in reality: heat only ever travels from a hot object to a cold object – never from a cold object to a hot object. So this tells us that the laws of physics which govern the behaviour of heat cannot be time-reversible. It also tells us that Newton's laws cannot describe the behaviour of heat, because – as we have just seen – Newton's laws are time-reversible.

As described earlier, Isaac Newton was perceived as almost a divine being in Britain during the Industrial Revolution. So it must have been a shock to discover that Newton's supposedly infallible laws – the laws which were treated with almost the same reverence as the Bible – were incapable of describing the behaviour of the new steam engines which were having such a great impact at that time.

The failure of Newton's laws to describe the behaviour of heat explains why there was such urgency in the early 1800s to develop a new science which was capable of describing the functioning of steam engines.

So let us now rejoin our story in France …

Reversibility and efficiency

Let us now return to consider the story of Sadi Carnot. If you remember, Carnot was obsessed with a desire to overcome the English technological superiority. To achieve his goal, Carnot aimed to improve the efficiency of French steam engines.

When we last left Carnot in the year 1824, Carnot had just published his book *Reflections on the Motive Power of Fire*. As described in the previous chapter, this was the book which launched the new science of thermodynamics – the science of heat and motion. And thermodynamics – unlike the laws of Newton – was the branch of physics which **was** capable of describing the behaviour of steam engines.

Also in his book, Carnot described how the principle of *reversibility* played a central role in determining the efficiency of a steam engine – or any machine. Why should reversibility be so important?

To understand why this is the case, consider the following thought experiment. Two ice hockey players are on a perfectly smooth surface of ice. As you can see in the following diagram, the player on the left (Player A) is hitting the puck to the player on the right (Player B):

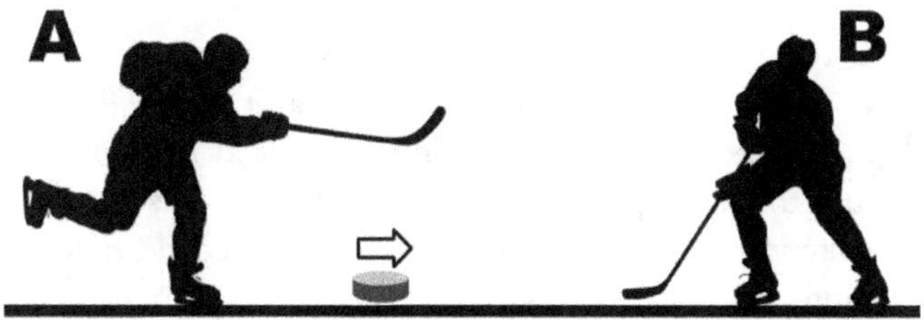

Let us imagine that Player A strikes the puck at a speed of 20 metres/sec towards player B. Let us also imagine that the ice is perfectly smooth and frictionless, so that the puck loses no speed on its journey. As a result, the puck arrives at Player B travelling at the same speed of 20 metres/sec.

Now let us imagine we make a movie of the sequence of events, just as we earlier made a movie of the snooker ball collisions. Here are three frames of the short movie which just shows the movement of the puck:

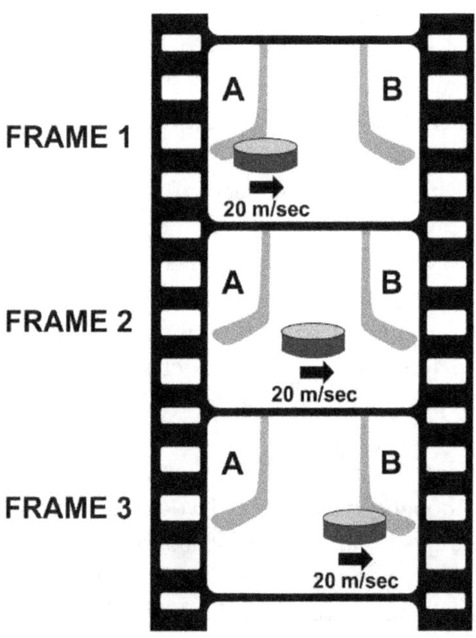

In the first frame of the movie, you can see the puck being struck by the hockey stick of Player A on the left, which results in the puck moving at a speed of 20 m/sec. Because the ice is perfectly smooth sind frictionless, there is no reduction in the speed of the puck in the second frame as it continues its journey across the ice. Finally, in the third frame, we see the puck hit the hockey stick of Player B – still travelling at a constant speed of 20 m/sec.

It is easy to see that if this movie was reversed, this sequence of events would still make sense. All that would change is that the direction of motion of the puck would be reversed, so the movie would show the puck moving in the opposite direction – at the same constant speed of 20 m/sec. For that reason, it is clear that this is a time-reversible situation.

Now let us consider the situation in which the playing surface of perfectly smooth ice is changed to be a surface of rough sandpaper. This would introduce the effect of friction on the motion of the puck. This would act to slow the puck on its journey across the ice. Although the puck would start its journey at the same speed of 20 m/sec, it might now slow down to a speed of just 5 m/sec by the time it reaches player B.

Here is the movie of this new situation. It can be seen that the puck loses speed as it moves across the sandpaper:

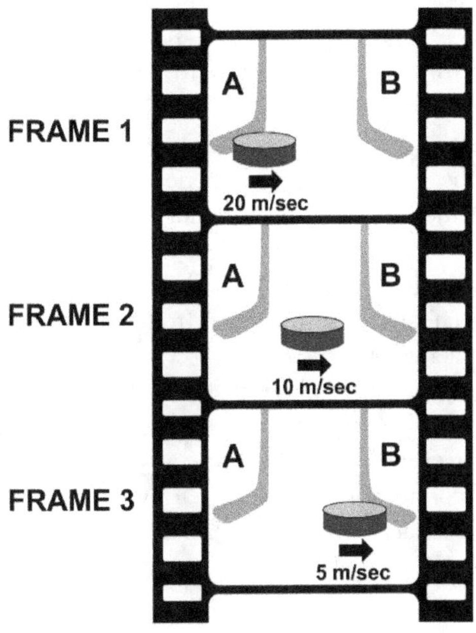

What has happened is that the puck has lost energy as it has passed across the ice. The energy associated with moving objects is *kinetic energy*, so the puck has lost kinetic energy as it moves across the sandpaper, due to the effect of friction.

Now let us imagine watching this new movie in reverse. In this new time-reversed movie, we would see the puck initially travelling at a speed of 5 m/sec as it leaves Player B, but speeding up as it moves across the sandpaper until it reaches a speed of 20 m/sec at Player A. This would, of course, never happen in reality: objects do not speed up as they travel across rough surfaces – they lose energy and slow down. So, because this time-reversed situation could never happen in reality, we can say that this has become an irreversible system.

So why has the previously reversible system now become an irreversible system? The cause of this irreversibility is the puck losing kinetic energy due to friction as it moved across the sandpaper. But the law of conservation of energy says that energy cannot be created or destroyed – it can only change its form. So the energy must have gone somewhere. But where did it go?

The answer is revealed when we realise that friction produces heat, a feeling we experience when we rub our hands together and feel them getting warm. Heat is a form of energy, so the kinetic energy lost by the puck is converted into heat energy by friction from the rough sandpaper. And that heat energy then transfers from the puck into the general cooler environment.

So the root cause of the irreversibility comes from the heat released by friction. Any friction in a system will introduce irreversibility.

This connection between irreversibility and the production of heat will play a central role later in this book when we come to consider consciousness.

The Carnot cycle

Sadi Carnot realised that any system which was losing energy due to friction would exhibit irreversible behaviour. And that energy loss would have the effect of reducing the efficiency of an engine. Therefore, in his book, *Reflections on the Motive Power of Fire*, Sadi Carnot explained that an engine with the greatest possible efficiency would be perfectly reversible. We now call this perfectly reversible engine a *Carnot engine*.

In order to achieve maximum efficiency, each individual step in the cycle performed by the engine must also be reversible. The complete sequence of reversible steps which is performed by the engine is called the *Carnot cycle*. In theory, the cycle could be interrupted at any point and the engine could then be operated in reverse.

So Sadi Carnot had finally achieved his goal of specifying how a perfect steam engine could be made: it should be reversible, or as near reversible as possible.

However, in reality, it can never be possible to make a perfectly reversible engine as it can never be possible to remove all loss-making factors such as friction. But, even for modern engineers, the goal of perfect reversibility is used to guide the process of efficient engine design.

A musical interlude

We have now come to the end of this chapter on reversibility. In the later chapters of this book, we will be returning to consider irreversibility. It will be seen that irreversibility potentially plays a key role in the emergence of consciousness.

However, right now, I think you deserve a reward. So I have specially written and recorded a song and music video based on the theme of this chapter.

The song is called *Turn Back Time*, and it is about irreversibility. The music video includes several examples of irreversibility including a collapsing house of cards, a breaking wine glass, and me getting wet (or, more accurately, me getting dry). Here is a link to the music video:

http://tinyurl.com/irreversibility

The song and video were influenced by the time-reversal movie *Tenet*.

Hopefully the song will have the effect of emphasizing the central role which irreversibility will be playing in this book.

I hope you enjoy the video!

4

HEAT, TEMPERATURE, AND ATOMS

As we have just seen, Sadi Carnot made great advances in understanding the behaviour of steam engines. However, Sadi Carnot still did not understand the true nature of heat. As explained in Chapter Two, Carnot still believed in the caloric theory of heat which proposed that heat was an actual substance. Carnot was fortunate that this mistaken theory of heat did not prevent his progress in understanding the steam engine. As Ingo Müller says in his book *A History of Thermodynamics*: "It was quite irrelevant whether he knew what heat was – and he didn't!"

A deeper understanding of heat was desperately required. In this chapter, we will see how that accurate, deeper understanding of heat was obtained. We will see how an entirely new theory had to be developed, a theory which relied on a completely new model of reality at its smallest scales.

At the end of this chapter we will also see that the theory of heat has implications for a theory of consciousness.

The rings of Saturn

The planet Saturn is the second-largest of the eight planets in the Solar System, almost a hundred times larger than Earth. The planet is mostly made of hydrogen and helium, with a total density less than water. But, of course, Saturn is most famous for its spectacular system of rings:

The rings around Saturn were first observed by Galileo Galilei in 1610 using his telescope. However, because of the low magnification of Galileo's telescope, the internal structure of the rings was unobservable. The true nature of the rings remained unknown. Were they connected to the planet, or separate? Were they solid, or made of small rocks, or even a fluid?

In 1857, in the hope of finally solving the problem, St. John's College in Cambridge announced a competition. A prize would be awarded to anyone who could solve the problem of the composition of the rings. The competition attracted the attention of 25-year-old James Clerk Maxwell, a highly-skilled mathematician who was a professor at Aberdeen University.

In 1859, after studying the problem for two years, Maxwell proposed his solution. In his essay *On the Stability of the Motion of Saturn's Rings*, Maxwell was able to show that if the rings were completely solid then the internal stresses would break the rings into smaller pieces as they orbited the planet. The same result would occur if the rings were made from a fluid: tidal waves would be created of such a monstrous size that the rings would eventually be torn apart.

Maxwell announced that the rings must be composed of a swarm of smaller objects, such as small rocks or lumps of ice. Occasionally, Maxwell realised, those rocks would collide with each other. This prediction was eventually confirmed by the images sent back from the Voyager spacecraft in 1980. The images showed that the rings are made of lumps of ice ranging in size from a radius of one centimetre to ten metres.

The Astronomer Royal described Maxwell's work as "One of the most remarkable applications of mathematics to physics that I have ever seen." James Clerk Maxwell won the £130 prize, and the physics community became aware of his name as a new rising star in the field.

Maxwell realised that his work on the rings of Saturn had wider applicability. Maxwell was one of a growing number of physicists who believed that all materials were made of smaller particles which were called atoms. Just as Maxwell had shown that the rings of Saturn were made from a huge number of smaller lumps colliding with each other, could he now use similar techniques to explain the properties of gases made of trillions of microscopic swarming atoms?

The kinetic theory of gases

If a gas really was made of trillions of atoms, then the challenge was to explain how the large-scale properties of a gas – such as its temperature and pressure – could be explained in terms of the interactions of the microscopic atoms in that gas. The solution was developed by Maxwell, together with the German physicist Rudolf Clausius and the

Austrian physicist Ludwig Boltzmann. Together, those three giants of 19th century physics created what we now know as the *kinetic theory of gases*.

The kinetic theory of gases models the atoms of a gas as microscopically small balls. Those balls move at high speeds, and, because of the vast number of atoms in a gas, collisions are extremely frequent. When these balls collide, the kinetic theory proposes that the balls bounce off each other perfectly elastically. As a result of these collisions, the motion of the atoms is completely random.

To provide some actual numerical values, the theory can be used to calculate that one molecule of oxygen in a room full of oxygen gas will be moving on average at the astonishingly high speed of 461 metres per second. When this result was announced, it raised some objections to the theory. As an example, consider the situation in which you are standing at one end of a room and someone wearing heavy perfume enters the other end of the room. With gas molecules predicted by the theory to travel at several hundred metres per second, it might be imagined that you would smell the perfume as soon as you see the person. But this is obviously not what happens – it takes several seconds for the scent to drift across the room.

The answer to this objection is that the atoms of the scent do not fly in straight lines. Instead, those atoms collide with other gas atoms as they zigzag across the room – just as the lumps of ice in the rings of Saturn collide with each other as they orbit the planet. This is the reason why the smell of the perfume takes several seconds to cross the room.

The kinetic theory of gases reveals that these collisions are occurring with extraordinary frequency. At room temperature, one oxygen molecule can travel an average distance of only 67 nanometres before it collides with another oxygen molecule, a distance which is 1,500 times smaller than the width of a human hair. Remembering that molecules are travelling at great speed, that means that a molecule will collide with another molecule 6.8 billion times per second. It is clear that the microscopic world is a world of extreme dynamism and constant, violent collisions.

The kinetic theory of gases provides an explanation of how large scale, measurable properties of a gas are produced by the interactions of the microscopic atoms in the gases. One such example is an explanation of how the pressure of a gas results from the forces applied by its atoms.

To see how this occurs, we might consider a sample of gas which is trapped in a cylinder, as shown in the following diagram. The atoms of the gas are drawn as small balls, bouncing off each other at high speed. The motions of the atoms are indicated by the small arrows. It can be seen that the cylinder contains a piston which is pushing down on the gas:

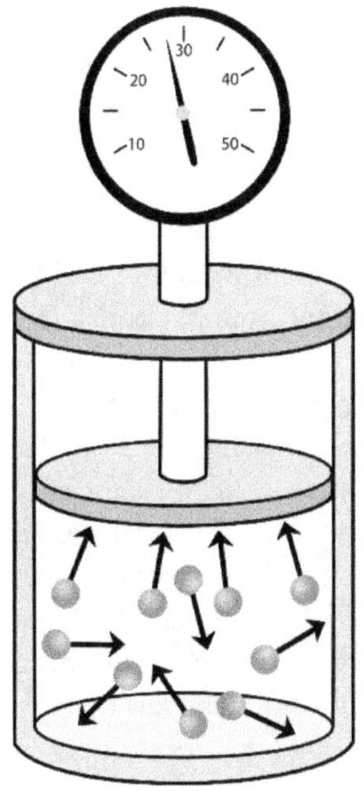

As has just been explained, the atoms of the gas are travelling at tremendous speed. Those atoms would strike the piston extremely fast, and this would result in a force being applied to the piston – like tennis balls striking a tennis racket. Because there are so many atoms in the gas, and they strike the piston with such high frequency, this would have the effect of exerting a constant outward force onto the piston. We would interpret this force as the pressure of the gas, and we could measure that pressure using a gauge (as shown in the previous diagram).

Hence it can be seen how the kinetic theory of gases can explain how a large scale, measureable property of the gas (such as pressure) can arise from the interactions of the microscopic atoms within that gas. The field of physics which considers how measureable properties arise from the interactions of microscopic particles is called *statistical mechanics*.

Heat

We might wonder how the kinetic theory of gases can explain the heat and temperature of a gas. Heat and temperature are closely related, but they are different properties. Let us consider heat first.

It was explained in Chapter Two that it was originally believed that heat was a substance, named either "phlogiston" or "caloric". It was suggested that caloric was a fluid, and caloric could be neither created nor destroyed. However, the main flaw in this theory is fairly obvious: if you create heat by friction – by rubbing two surfaces together – then you can create an endless source of heat. It was clear that heat could not be a substance which was limited in amount.

In his 1857 paper *On the Nature of the Motion which we call Heat*, Rudolf Clausius rejected the caloric theory of heat. Instead, Clausius suggested that heat was motion: the motion of atoms in an object. This makes a lot of sense, after all, when you rub your hands together they get hot. In that case, the motion of your hands is causing the atoms of your hands to move, which generates heat.

Clausius realised that each of those moving atoms would possess an amount of kinetic energy, the energy associated with movement. Let us consider how atoms and molecules can move. In a gas, three types of molecular motion are possible, as shown in the following diagram:

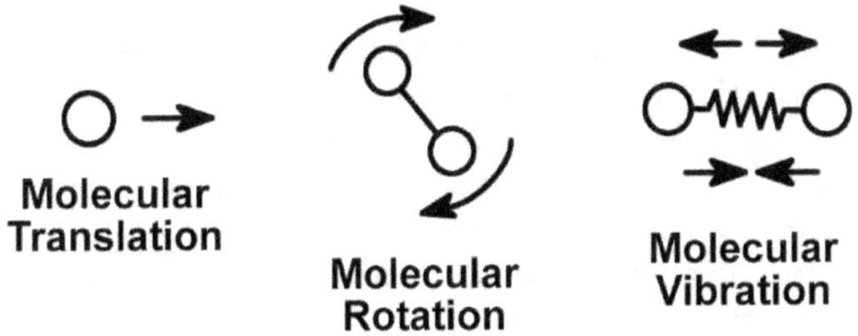

As the previous diagram shows, the three possible types of molecular motion are translation (movement in a straight line), rotation, and vibration. Each different way in which the molecule can move is called a *degree of freedom*. There is some kinetic energy – the energy of motion – associated with each degree of freedom. In other words, each different way in which the molecule can move adds an amount of energy to the total energy of the molecule. If we then add up the total energy of all the molecules we get the total thermal energy (the heat) of the gas.

Heat, therefore, is the energy associated with the random motion of atoms and molecules in an object.

Temperature

So the total amount of thermal energy in an object is the heat in that object. Let us now move on to consider temperature – which is different from heat.

We can measure temperature using a thermometer – or even our bare hands. In terms of physics, the temperature is the **average** kinetic energy of the atoms in an object. The word "average" is crucial here. A very large object might contain a great amount of heat energy in total, but – because of the size of the object – the average temperature might be surprisingly low. An example of this is the temperature of the sea. The sea contains a huge amount of heat energy in total, but, because of its size, the temperature of the sea when we put our hand in the water – the **average** amount of energy in a small sample – can be chilly!

The ocean has absorbed more than 90% of the heat caused by human activity. John Abraham, a professor of thermal sciences at the University of St. Thomas in Minnesota has explained how much heat energy is absorbed by the ocean: "It's about five Hiroshima bombs of heat, every second, day and night, 365 days a year". This reveals the vital role of the oceans in absorbing heat in the battle against global warming.

So the heat energy contained within the ocean is huge – but if you dip your toe into the sea, the water will feel cold! This is an example of the difference between temperature and heat.

The bridge between the human world
and the atomic world

Let us now consider in more detail the connection between temperature and the properties of individual atoms.

The kinetic theory of gases can be used to calculate the amount of kinetic energy of a single molecule of a gas. That kinetic energy can be found from the following formula:

$$\text{Kinetic energy of molecule} = \frac{3}{2}kT$$

where T is the temperature of the gas and k is a constant value known as *Boltzmann's constant.*

But where does the 3/2 factor come from? Well, as explained earlier in this chapter, gas molecules have various degrees of freedom, which means they have various directions in which they are free to move. Each degree of freedom adds an amount of kinetic energy to the molecule which is equal to:

$$\frac{1}{2}kT$$

With three dimensions of space in which the molecule is free to move (three degrees of freedom), that means that this amount of energy gets added three times to the total energy of the molecule, thereby resulting in the 3/2 factor shown earlier.

Now let us consider how we measure temperature. If I want to measure temperature, I would use a thermometer. A thermometer is a fairly large instrument which I can hold in my hand. A thermometer is definitely part of the human-scale world. Because of its size, we would say that a thermometer is a *macroscopic* object (a human-scale object).

In contrast, I cannot see an atom because of its incredibly small size: it is unobservable. I cannot hold a single atom in my hand, or use a single atom as a tool. An atom is not part of my human world. Because of its small size, we would say that an atom is a *microscopic* object (as opposed to macroscopic).

But the analysis which has just been described managed to connect temperature with the kinetic energy of an individual molecule. How was that magic achieved? Let us again consider the previous formula for the kinetic energy of a molecule:

$$\text{Kinetic energy of molecule} = \frac{3}{2}kT$$

We can see that the factor which links temperature of a gas with the energy of one of its molecules is Boltzmann's constant, k. For this reason, it is often stated that Boltzmann's constant represents "the bridge between the microscopic and macroscopic world". For an example, see the following page on the Boltzmann.com website:

http://tinyurl.com/boltzmannconstant

As that webpage states:

> *Boltzmann's constant bridges macroscopic physics (temperature, entropy) with microscopic physics (average energy of particles/atoms/molecules).*

Consciousness is part of the human experience, part of the human world. So consciousness is a *macroscopic* phenomenon. Whereas the motion of particles in our brains – out of which consciousness emerges – is a *microscopic* phenomenon.

This point was made by the physicist Carlo Rovelli in a recent YouTube video (**http://tinyurl.com/rovellivideo**). In the video, Rovelli attempts to understand consciousness within the known laws of physics, and emphasises the macroscopic nature of consciousness. According to Rovelli:

> *Consciousness is a macroscopic phenomenon. Subjective experience ... is a macroscopic phenomenon.*

The challenge for physicists is to explain how consciousness emerges from the trillions of interacting particles in our brains. How can we link microscopic activity to a macroscopic phenomenon? For this reason, I would suggest that Boltzmann's constant will likely play a central role in any final theory of consciousness. The theory of consciousness presented later in this book will have a central role for Boltzmann's constant.

Consciousness and physics

This line of reasoning, I would suggest, provides us with the first indication of why thermodynamics is the branch of physics which is best suited for studying the phenomenon of consciousness.

When we consider a brain, we observe a staggeringly complex swarming of microscopic particles. Somehow, our consciousness emerges as a macroscopic phenomenon from the interactions of these particles in the same way that the temperature or pressure of a gas arises from interactions between the atoms of that gas. Consciousness is, therefore, a phenomenon which needs explaining in terms of the activity of its microscopic constituents, in the same way that explanations were found for temperature and pressure in terms of microscopic atoms – as this chapter has explained.

Although, historically, there has been little interest in consciousness from physicists, several high-profile physicists now appear to be taking the problem seriously. And, I am pleased to say, I detect a general focus on thermodynamics as a likely solution.

This trend towards considering thermodynamics as a solution was clearly shown in the 15th February 2020 cover story of *New Scientist* magazine. The article was called *Finding Our Place in the Universe* and was written by Richard Webb. In the article, Webb interviewed several prominent physicists including Sean Carroll and Carlo Rovelli. In the following extended extract from the article, the possible explanation for consciousness in terms of thermodynamics is explored, with particular reference to the kinetic theory of gases:

For Carroll and many others, the answer lies not in mysticism, but in emergence. This is the idea that behaviours and properties that are inscrutable when you look at single components of a complex system pop into existence when you view things as a whole. There are plenty of precedents. The temperature or density of a gas, for example, doesn't mean much at the level of single molecules. Look at all the molecules of the gas together, however, and they are measurable quantities that explain physical change: how temperature differences cause heat flows, for example, or how a gas pushes a piston when compressed.

The catch is that human brains are phenomenally complex. A fully grown one contains nearly 100 billion interconnected neurons, and we are still far from establishing how they produce the felt states of our conscious world, let alone how those states interface with the wider world: "It's one thing to say I can explain the temperature and density of the air by hypothesising that it is made of molecules bumping into one another" says Carroll. "It seems quite a more dramatic project to say I can explain mind and choices and consciousness as emerging out of atoms and molecules bumping into one another."

For physicists, the challenge is clear. An explanation of the macroscopic phenomenon of consciousness in terms of microscopic atoms is required – just as explanations were found for temperature and pressure.

With that in mind, let us continue our story …

5

ENTROPY

When we last left Sadi Carnot it was the year 1824, and Carnot had just published *Reflections on the Motive Power of Fire*. The principles which Carnot described in that book still influence engineers to this day. Unfortunately, Carnot did not live to see the widespread adoption of his ideas as he tragically died in Paris during a cholera epidemic in 1832 at the age of just 36.

Carnot's work remained largely unnoticed for almost twenty years after his death before it was rediscovered by a few notable scientists. That small group included the German physicist and mathematician Rudolf Clausius.

It was Clausius who, in the 1850s and 1860s, analysed the work of Carnot and placed it on a stronger theoretical foundation. Clausius proceeded to make a series of theoretical breakthroughs, in the process becoming one of the key founders of the science of thermodynamics.

Despite Carnot's advances, one great mystery remained about the behaviour of heat. That mystery was the tendency of heat to travel in only one direction: from hot to cold. As explained in Chapter Three, this was a mystery because Newton's laws – and the fundamental laws of physics in general – are all reversible in time. It was Clausius who provided the breakthrough in solving the problem when he introduced the concept of *entropy*.

In developing the concept of entropy, Clausius carefully analysed the writings of Sadi Carnot about steam engine efficiency. Clausius considered a system which had heat energy added to it. Clausius then defined the increase in the entropy of that system as follows:

$$\text{Entropy change} = \frac{\text{Heat}}{\text{Temperature}}$$

Crucially, Clausius also combined this definition of entropy with a rule to describe its behaviour:

The entropy of a closed system always increases in time.

This rule is called the *second law of thermodynamics*, and it is considered to be one of the most important and unbreakable rules in physics.

It can immediately be seen that there is a crucial difference between this law and the laws of Newton. While the laws of Newton do not specify any particular direction of time, it can be seen that the second law of thermodynamics says that entropy always increases in the forward direction of time – never in the reverse direction of time. So, unlike the laws of Newton, **a direction of time is explicitly stated within this law.** It is clear that we are moving away from the laws of Newton and into a new world of physics.

An example calculation of entropy

Let us now consider an example calculation of entropy, according to Clausius's definition. The following example is from the excellent free online physics textbook provided by OpenStax. Here is a link to the relevant page in the book:

http://tinyurl.com/entropybook

The example described in the book considers the following scenario. Imagine a metal bar. The left half of the bar is at a very high temperature of 600 Kelvin (327 degrees Celsius), while the right half of the bar is at a low temperature of 250 Kelvin (-23 degrees Celsius). This arrangement is shown in the following diagram, with one half of the metal bar being hot, and the other half being cold:

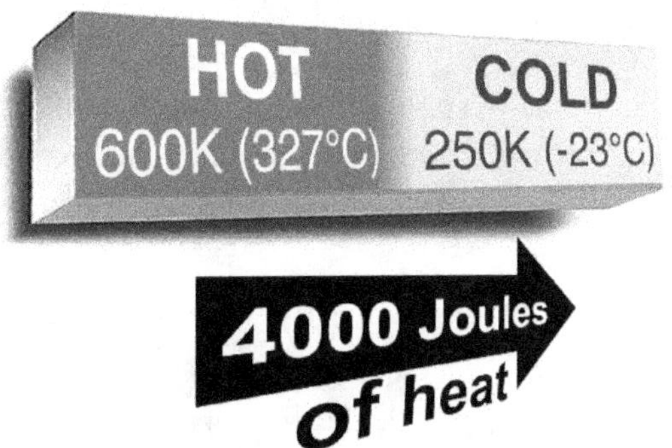

If we leave the bar for a period of time, we will find that heat energy naturally flows from the hot end of the bar to the cold end of the bar (heat naturally flows from hot to cold). Let us consider the situation in which 4000 Joules of heat energy flows from hot to cold (shown by the arrow in the previous diagram). We then ask the question: what is the overall change in the entropy of the entire system?

Firstly, let us remind ourselves of Clausius's definition of the change in entropy:

$$\text{Entropy change} = \frac{\text{Heat}}{\text{Temperature}}$$

Now let us calculate the entropy change in the hot, left half of the metal rod:

$$\text{Entropy change} = \frac{-4000}{600} = -6.67 \text{ J/K}$$

Note that this is a negative value, representing a decrease in entropy. But that is strange as it was just explained that entropy always increases. So what is going on?

All becomes clear when we consider the cold, right half of the metal rod:

$$\text{Entropy change} = \frac{4000}{250} = 16 \text{ J/K}$$

This time the result is positive, representing a large increase in entropy. If we then consider the total entropy change in both halves of the metal rod, we get the following value:

$$\text{Total entropy change} = 16 \text{ J/K} - 6.67 \text{ J/K} = 9.33 \text{ J/K}$$

which is a positive value. So the total change in entropy of the rod is a positive value. In other words, the total entropy has increased – in agreement with the second law of thermodynamics.

This calculation has revealed how the second law of thermodynamics – the rule that entropy always increases – can describe the flow of heat from hot to cold.

However, for many years after Clausius's discovery, no one actually knew what entropy was. Why should there be this strange property which was always increasing?

The answer was provided by an Austrian physicist, and it was an answer which changed our view of reality …

The atoms of Ludwig Boltzmann

Ludwig Boltzmann was born in Vienna in 1844. In 1866 he obtained his doctorate from the University of Vienna, and by 1869 he had become a professor of mathematical physics at the University of Graz.

It was when he was a student at the University of Vienna that Boltzmann performed his first important work. That work was on the kinetic theory of gases, which was considered in the previous chapter. If you remember, the kinetic theory of gases revealed how the temperature of a gas represented the average kinetic energy of a single molecule of that gas. It was explained how the factor which links the overall temperature of the gas to the energy of a single molecule is now called Boltzmann's constant, named in recognition of Ludwig Boltzmann's pioneering work.

Here is a photograph of Ludwig Boltzmann:

It was clear, then, that Ludwig Boltzmann believed in the reality of atoms, that a gas was composed of trillions of atoms moving around with considerable kinetic energy. What is more, Boltzmann realised that if a gas was, indeed, made of atoms then a simple explanation for entropy emerged. Remember, the second law of thermodynamics stated that entropy had to always increase. Boltzmann realised that this made sense if entropy represented the amount of disorder of the atoms in a gas. If that was the case, then entropy would naturally tend to increase – in the same way that a set of playing cards would tend to become more disordered when it was shuffled.

It really is quite a simple idea. But, as we shall now see, this simple idea was capable of explaining the flow of heat from hot to cold.

Let us imagine the metal rod which was described in the previous section. The left half of the rod is hot, while the right half of the rod is cold. We might consider this arrangement as representing a type of "ordering" of the atoms in the rod: the faster-moving atoms (hotter) have been ordered so that they are in the left half of the rod, while the slower-moving atoms (colder) have been ordered so that they are in the right half of the rod.

Over time, as the atoms in the two halves of the rod interact with each other, this ordering would tend to disappear. As the atoms in the two halves randomly collide with each other, some of the atoms in the left half of the rod would slow down, while some of the atoms in the right half of the rod would speed up.

Eventually, an equilibrium position would be reached in which the atoms in both halves of the rod have the same average velocity. At this point, the entire rod is at the same temperature. All of the initial ordering of the atoms would have disappeared. This would be the point of maximum disorder: maximum entropy.

But how would we interpret this increase of disorder? We would observe the cold half of the rod getting warmer, while the hot half of the rod becomes colder. So we would interpret the increase of entropy as a flow of heat from hot to cold. This atomic model of Boltzmann could therefore explain how an increase in entropy would lead to a flow of heat from hot to cold – but never in the reverse direction.

We now know that Boltzmann's theory was correct, and his explanation of entropy is accepted as one of the great achievements of 19th century physics. However, the theory received a great deal of opposition at the time. All of Boltzmann's work depended on the existence of atoms. However, when Boltzmann proposed his theory, the existence of atoms was by no means certain. Indeed, many German scientists believed that atoms did not exist, and they were openly critical of Boltzmann's idea. Influential critics such as the Austrian physicist Ernst Mach subscribed to the positivist school of thought which believed that physics should only consider things which can be directly observed – and atoms are, of course, too small to be directly observed.

Throughout his life, Ludwig Boltzmann had suffered from manic depression. This continued rejection of his ideas contributed to the slow decline in Boltzmann's mental state.

In September 1906, Boltzmann took a vacation to the coastal village of Duino, near Trieste, with his wife and daughter. The trip was specifically intended to improve his state of mind. Indeed, by the end of the holiday his health seemed much improved. About six o'clock in the evening of September 5th, Boltzmann's wife and daughter went to the beach for a swim, and Boltzmann assured them he would join them shortly. But Boltzmann never arrived. Tragically, Ludwig Boltzmann had hanged himself from a window in his hotel room.

Order and disorder

The interpretation of entropy as a form of disorder provides an intuitive notion of the nature of entropy. However, it also introduces a problem. Just saying that one arrangement of particles is "ordered" while a different arrangement is "disordered" appears to be a matter of personal taste. Who gets to decide what is an ordered state and what is a disordered state?

As an example, if we look into the bedroom of a teenager, it might look to be in a disordered state – a complete mess. Take a look at the following image, for example. You might think it looks completely disordered. But maybe the teenager would insist that all of the items are positioned just where he or she wants those items to be, and so, according to the teenager, the room is perfectly ordered.

So what is the answer? Does the following image show an ordered room or a disordered room? Hence, is it low in entropy or high in entropy?

Ordered or disordered?

It is clear that if we want to use the model of entropy as disorder then we need to find an unambiguous definition of the difference between order and disorder. In order to solve that problem, we first need to introduce the concepts of *microstates* and *macrostates*.

Firstly, remember from the discussion in the previous chapter that the prefix "micro" refers to very small objects, usually atoms, while the prefix "macro" refers to the large human-scale world. So the term "microstate" refers to the state of a system at the smallest level: the positions and velocities of its atoms. However, humans cannot see atoms, so the term "macrostate" refers to the state of a system as recognisable by a human.

A classic example of a macrostate is the temperature of a system. Temperature is a macrostate as it can be measured by a human-scale device: a thermometer. If an object is at a certain temperature, that means that the atoms of the object are moving randomly, possessing a certain kinetic energy. If the atoms then move to different positions, that would mean that the system is then in a different microstate as the microscopic positions of the atoms would have changed. However, the temperature of the system might well be the same as before. So the temperature macrostate of an object is associated with a large number of different microstates (different arrangements of atoms).

Let us consider the example of the teenager's room again. If the room is in a state of "order", then there are not many associated microstates. The reason why that is the case is described by Peter Hoffman in his book *Life's Ratchet*:

> *There are only a few arrangements (microstates) compatible with a tidy room: clothes in the closet, grouped by long-sleeve shirts, short-sleeve shirts, slacks, jeans, and so on; books on the shelves, sorted alphabetically by author; and so forth.*

So when a room is in an ordered state, there are not many different possible arrangements of the objects in that room – not many microstates. However, when a room is in a disordered state, there are a vast number of possible different arrangements. As Peter Hoffman again describes:

There are almost unlimited ways the room could be messy: jeans on top of the computer, books on the floor, shirts strewn over the bed.

So this provides us with a way of distinguishing between ordered and disordered states. A disordered state will have many more microstates associated with it than an ordered state will have. A disordered room has many more ways that it can be disordered. This allows us to categorically state – for the reason described by Peter Hoffman – that the teenager's room in the previous diagram is in a state of disorder!

And, as entropy is defined as being the amount of disorder in a system, this allows us to create a formula for the amount of entropy in a system. The amount of entropy of a system will be proportional to the number of microstates associated with the current macrostate of that system: more microstates means more entropy. This allowed Ludwig Boltzmann to develop a formula for the entropy of a system.

Here is Boltzmann's formula:

$$\text{Entropy} = k \ln W$$

where W represents the number of microstates associated with the current macrostate of the system, k represents Boltzmann's constant (introduced in the previous chapter), and "ln" represents the natural logarithm function. The units of entropy are joules per kelvin.

Put simply, we see that the formula tells us that the greater the number of microstates associated with the current macrostate, the greater will be the entropy of that macrostate.

Boltzmann's formula for entropy was such a tremendous achievement that you can see the formula engraved near the top of Boltzmann's magnificent tombstone in Vienna:

6

MAXWELL'S DEMON

James Clerk Maxwell was born in Edinburgh, Scotland, in 1831. Maxwell was born into a wealthy and eccentric family. As an example of the adventurous and eccentric nature of the family, Maxwell's Scottish grandfather, James Clerk, was a captain in the East India Navy. Unfortunately, his ship was wrecked in the River Ganges. James Clerk managed to avoid drowning by floating to shore on his inflated bagpipes. When he reached the shore, he played the bagpipes to scare away some predatory tigers.

James Clerk Maxwell inherited his grandfather's adventurous and curious nature. As a young boy, he would often ask questions about the operation of the devices he found in the house. In his Scottish accent he would ask: "What's the go o' that?". If he did not find the answer to his question to be satisfactory, he would then ask: "But what's the *particular* go o' that?"

Maxwell's talents as a superb mathematician became clear when he studied as an undergraduate at the University of Edinburgh and Cambridge University. It was while he was in Cambridge that he started his work on analysing the composition of Saturn's rings. As described in Chapter Four of this book, Maxwell realised that the rings had to be composed of many smaller rocks which would frequently collide with each other. Also as described in Chapter Four, this work on the interactions of rocks in the rings of Saturn would lead to James Clerk Maxwell becoming one of the most important contributors to the kinetic theory of gases.

However, Maxwell's work on the kinetic theory of gases is not the work for which James Clerk Maxwell is most famous. After securing a professorship at Kings College in London, Maxwell started to analyse the connection between the electric force and the magnetic force. In his 1865 paper *A Dynamical Theory of the Electromagnetic Field*, Maxwell unified electricity and magnetism, showing that they were two aspects of a single electromagnetic force. As a result of that unification, Maxwell had the historic insight that light must be an electromagnetic wave.

Mainly for his achievement in electromagnetism, Maxwell is now regarded as one of the most important physicists of all time. When *Physics World* magazine conducted a poll of 100 leading physicists to discover who was considered to be the greatest physicist of all time, James Clerk Maxwell came third, only behind Albert Einstein and Isaac Newton.

Here is a drawing of James Clerk Maxwell:

Maxwell enjoyed a lifelong correspondence with his good friend Peter Tait. Tait was an outstanding Scottish physicist and an expert on thermodynamics. They would send letters to each other, often containing humorous stories, or observations about science.

In December 1867, Maxwell had an idea which he found deeply puzzling, so he described the idea in a letter to Tait. The idea contained in Maxwell's letter was so startling that it is still having ramifications to this day. Let us consider Maxwell's remarkable idea.

As has been described, Maxwell was a major contributor to the kinetic theory of gases, and he was well aware of the latest developments in thermodynamics. In his letter to Tait, Maxwell described a thought experiment which apparently broke the "unbreakable" second law of thermodynamics.

If you remember from the previous chapter, the second law of thermodynamics states that the entropy of a closed system will always increase over time. If we consider entropy to be the amount of disorder in a system, then this makes intuitive sense: ordered systems tend to become more disordered over time.

However, in his letter to Peter Tait, James Clerk Maxwell described the following scenario. Imagine a sealed box which contains a gas. The box is divided into two compartments by a solid partition in the middle of the box. The gas is initially at the same temperature in both compartments.

There is a small trapdoor in the partition which can be opened and closed to allow individual atoms of the gas to pass from one compartment to the other.

Maxwell then asked us to imagine a so-called "demon" who was in charge of opening and closing the trapdoor. The demon is in charge of performing the following operation. If the demon sees a fast-moving atom of gas approaching the trapdoor from the right-hand compartment, the demon opens the trapdoor to allow the atom to cross to the left-hand compartment. And, vice versa, if the demon sees a slow-moving atom of gas approaching the door from the left-hand compartment, the demon opens the trapdoor to allow the atom to cross to the right-hand compartment.

The following diagram shows the demon opening the trapdoor to allow a fast-moving atom of gas to cross from the right-hand compartment to the left-hand compartment:

We see that, over time, the effect of the demon's actions will be that the left-hand compartment contains faster-moving atoms than the right-hand compartment. As heat is nothing more than the movement of atoms, that means that the left-hand compartment will become hotter than the right-hand compartment.

And so this is where the puzzle emerges. As described in Chapter Two of this book, a heat engine operates on the basis of a temperature difference: a "waterfall of heat". So we could connect a heat engine in between the two compartments and use that temperature difference to perform work. In theory, this could continue forever as the demon continues to order the gas atoms according to their speed. It appears we have found a form of perpetual motion machine, which is, of course, impossible.

But the puzzle which really concerned James Clerk Maxwell is that the demon appears to have broken the second law of thermodynamics. The second law states that entropy – disorder – of a system will always tend to increase. But the demon has ordered the gas molecules according to their speed, so the demon has ordered the atoms. As an ordered system has lower entropy, this has had the effect of reducing the entropy of the system. It would appear that the second law of thermodynamics has been broken.

This mystifying thought experiment is now called *Maxwell's Demon*.

In 2012, the BBC made an excellent two-part documentary series about thermodynamics called *Order and Disorder*, presented by Prof. Jim Al-Khalili. A link to the first episode was presented earlier in Chapter Two of this book. The second episode considered Maxwell's Demon. Here is a link to the second episode on YouTube:

http://tinyurl.com/videomaxwell

The whole video is excellent. The discussion of Maxwell's Demon is at the 23-minute mark of the video.

The physicist Paul Davies has recently written a book about information in biological systems. The book is titled *The Demon in the Machine*, and it contains a chapter on Maxwell's Demon. In that chapter, Paul Davies explains that the problem of how Maxwell's Demon could seemingly break the second law of thermodynamics remained unsolved for over sixty years. According to Paul Davies: "For many decades, it lay like an inconvenient truth at the core of physics, an ugly paradox that most scientists chose to ignore."

To find the solution, physicists had to look in a very dark and surprising place …

Inside the mind of the demon

In 1929, the Hungarian physicist Leo Szilard believed he had found a solution to Maxwell's Demon. Szilard considered the second law of thermodynamics, which says that the entropy of a system will always tend to increase. Szilard realised that, in the case of Maxwell's Demon, **the whole system also included the demon** – not just the gas in the box. If the entropy of the demon was increasing by a greater amount than the entropy of the gas in the box was decreasing, then the entropy of the system **as a whole** would be increasing. In that case, the second law of thermodynamics would not be violated after all.

But this posed a second mystery. Why should the entropy of the demon be increasing? At first glance, it appeared the demon was not doing very much at all – just opening and shutting the trapdoor. That surely did not require much expenditure of energy. It was even possible to imagine that the trapdoor was completely frictionless – in which case the demon would be performing no physical work at all!

But, of course, the demon was actually exerting a huge amount of effort – but it was not physical work. No, all the work was in the demon's mind. The demon was having to perform a superhuman feat in his mind, measuring the velocity of every single atom of the gas, and then calculating whether or not the trapdoor should be opened. When the vast number of atoms in the gas is considered, Szilard realised that the increased entropy generated by the superhuman mental processes of the demon could hold the key to solving the mystery.

In 1929, Szilard published this proposed solution to Maxwell's Demon in his paper *On Entropy Reduction in a Thermodynamic System by Interference by Intelligent Subjects*.

The use of the word "intelligent" in the paper's title is remarkable. Up to that point, physicists had only talked about simple, clear properties such as velocities and forces, properties which could be easily measured. But now, for the first time, Szilard was bringing "intelligence" within the domain of the laws of physics.

Physicists had always avoided the concept of intelligence because it was something internal to the human mind – it was not something that could be measured, or studied objectively. After all, we cannot see inside a mind.

But Leo Szilard had managed to see inside the mind of the demon.

Information, entropy, and bits

In the remainder of this chapter, let us consider Leo Szilard's solution of Maxwell's Demon in more detail, based on the argument which Szilard presented in his original paper.

The first step taken by Szilard was to simplify the problem. In order to do that, he considered a box which was filled with very little gas. In fact, he considered a box containing just a single atom of gas.

Initially, before the demon has had a chance to do his ordering job, we would consider the gas (at least, the single atom of the gas) to be in a "disordered" state. At that point, the single atom of the gas is flying around the entire box randomly, and at any particular moment in time it could be in one of two positions: either in the left-hand side of the box, or in the right-hand side of the box. The task of the demon is then to "order" the gas, which means using the trapdoor to trap the single atom in either the left-hand compartment or the right-hand compartment – depending on the speed of the atom.

So, in which side of the box should the demon trap the atom? In order to calculate that, the demon must first observe the atom, and measure its velocity. The demon must then use its mind to perform a simple comparison operation to determine if the particle is either slow or fast. On the basis of the result of that operation, the demon then either opens the trapdoor, or leaves the trapdoor closed.

We can therefore see that the end result of all the processing in the demon's mind (observing, measuring, calculating) is to produce just a single piece of **information**, and that single piece of information can take one of only two values: "slow" or "fast". The following diagram shows Leo Szilard's demon, with only a single atom of gas in the box, and the demon having to use its mind to calculate if the atom is "slow" or "fast":

A piece of information which can only take one of two possible values (in this case, "slow" or "fast") is called a *bit*. "Bit" is short for "binary digit". By convention, a bit can only take the value 0 or 1. We might therefore represent a slow atom by a bit with the value 0, and represent a fast atom by a bit with the value 1. The task which the demon must perform in its mind then becomes clearer:

With one atom in the box, the demon must generate one bit of information.

Of course, the demon does not have to calculate the information himself in his own mind. Instead, the calculation could be performed elsewhere by some third party, and then that information could be transmitted to the demon who could then use that information to decrease the entropy of the gas. This situation is shown in the following diagram. In this case, all the atoms of the gas are now in the box. The demon therefore needs a large number of bits of information – one bit per atom of gas. This information is calculated elsewhere, and is then transmitted to the demon as a long string of bits:

You can see in the previous diagram that the demon is now holding a smartphone. The demon receives the transmitted string of bits (zeroes and ones) on his smartphone, and then by reading the information one bit at a time he knows whether to open the trapdoor or leave it closed. Hence, the demon can use the transmitted information to lower the entropy of the entire gas.

So here we have revealed a very important principle: information is the inverse of entropy. If relevant information is available, then that information can be used to lower the entropy of a system. As Paul Davies states in his book *The Demon in the Machine*: "Information is, in some sense, the opposite of entropy."

And, vice versa, if information is lost, then the entropy of the system will increase. We will see that this second point will have major implications later in this book: if information is lost, then entropy will increase.

Back inside the mind of the demon

Let us now return to consider the situation in which the demon processes the information in his own mind. It is in this case that Leo Szilard had his moment of brilliant inspiration. As described earlier, Szilard realised that if the demon reduced the entropy of the gas, then the second law of thermodynamics would then require the entropy inside the demon's brain to increase by at least an equivalent amount.

This was a remarkable discovery. The brain is an incredibly complicated system, containing approximately 86 billion neurons, wired in an almost inconceivably complex manner. Similarly, the range of various thoughts which might be passing through the demon's mind might appear to be uncountably huge in number: "Do I open the trapdoor? Yes or nor? It is a fast-moving particle so, yes, I will open the trapdoor."

It might appear impossible that our current knowledge of physics could ever be used to analyse such a complicated system. But, via an indirect technique using the second law of thermodynamics, Szilard had shown how it could be done – at least in the Maxwell's Demon example. Szilard had shown that if we use our minds to process information and make a decision about whether or not to open the trapdoor, the entropy of our brain has to increase by a certain, fixed amount. In this example at least, Leo Szilard had brought the human brain and the workings of the human mind into the domain of exact physics.

So by how much, exactly, would the entropy of the brain have to increase? To answer that question, we have to use the formula of Ludwig Boltzmann which was described at the end of the previous chapter. If you remember, Boltzmann's formula for the entropy of a system was given as:

$$\text{Entropy} = k \ln W$$

where W represents the number of microstates associated with the current macrostate of the system, and k is Boltzmann's constant.

With that in mind, let us once again consider the situation in which there is only a single atom of gas in the box. In this situation, before the demon has performed his ordering operation, we might consider the gas to be in a "disordered" macrostate. The atom is flying around the entire box randomly, and at any particular moment in time it could be in one of two positions: either in the left-hand side of the box, or in the right-hand side of the box. Hence, this "disordered" macrostate is associated with two microstates (remember: microstates deal with the positions of individual particles, individual atoms).

If we then substitute the value of W equal to 2 in the previous equation, we then get the value for the entropy of the initial disordered system as:

$$\text{Entropy} = k \ln(2)$$

And this is therefore the amount by which the entropy of the gas is reduced when the demon performs his ordering operation on the single atom of the gas.

So, even though the brain is an inconceivably complex structure, we have arrived at a very precise result. The entropy of the demon's brain would have to increase by an amount equal to $k\ln2$. Leo Szilard really had brought the operation of the brain and mind into the domain of exact physics.

But this result raises another question. What is the underlying reason for this precise value of entropy increase in the brain? Where does this value of $k\ln2$ originate in the processing of information in the brain? In the next two chapters, we will discover the solution to that question.

And we will see that the explanation is entirely logical …

7

LOGIC

The development of logic dates back to antiquity. In ancient Greece, philosophers such as Aristotle and Plato would often spend their days having arguments about mathematics and philosophy. These arguments would sometimes extend for many days, or remain permanently unresolved. Aristotle believed that it should be possible to develop a clear method for resolving these arguments. How could the truth or falsity of an argument be unambiguously established?

The methods of logic were developed for precisely this reason of settling arguments. It was Aristotle who suggested the first method for determining whether a claim or statement made by one of his fellow philosophers was true or false. With a clear method in place, the philosophers on both sides of the argument would be forced to accept the conclusion.

To understand the basis of Aristotle's approach, it must first be explained that at the core of logic lies statements – or claims – which can be either true of false. No other value is possible. These statements are called *propositions*.

As an example, "The cat is on the mat" is a proposition as it is a statement which can be either true or false. The following image shows how the proposition "The cat is on the mat" could be evaluated to one of two different values:

At this point, you might recognise the similarity between a proposition and a "bit" of information (considered in the previous chapter), as a bit is also restricted to taking only one of two possible values: 0 or 1. We will shortly be returning to consider this similarity.

Aristotle's method of logic was based on stating some initial assumptions, called *premises*. Those premises would be simple propositions which everybody could agree were true. For example, "Socrates is a man" is an example of a premise – a simple statement which everyone can agree is true.

In order to create a convincing *argument*, those initial premises would then be connected to produce a single *conclusion*. Here is a famous example of one of Aristotle's arguments which consists of two premises (the first two lines) and a single conclusion (the last line):

> *Socrates is a man.*
> *All men are mortal.*
> *Therefore, Socrates is mortal.*

It can be seen in this example that the conclusion ("Socrates is mortal") logically follows from the two premises (the previous two lines).

It is important to note that there may be a large number of premises, but **there is only ever one single conclusion.**

This crucial point is emphasized by Dr. Siu-Fan Lee in his book *Logic: A Complete Introduction*:

> *An argument contains one and only one conclusion. If we want to make more than one conclusion, then there must be more than one argument.*

Truth tables

The example of Aristotle's argument included two premises ("Socrates is a man" and "All men are mortal"). However, there is no upper limit on the number of premises. Multiple premises can be connected by using *logical operators*.

Logical operators such as AND and OR operate in a similar manner to how they operate in the English language. The following example shows two premises connected by the AND operator:

IF
  I have a knife
AND
  I have a fork
THEN
  I can eat the meal

In this example, the conclusion ("I can eat the meal") is only true if both the first premise **AND** the second premise are true, and is false otherwise. The premises are therefore connected by the AND logical operator. The use of logical operators plays a central role in *Boolean logic*.

Boolean logic is a branch of logic which resembles mathematical arithmetic. In Boolean logic, the logical value TRUE is replaced by the number 1, and the logical value FALSE is replaced by the number 0. In that case, if a premise is true then it is given the value 1, and if it is false then it is given the value 0.

In arithmetic, numbers are connected using operators such as addition and multiplication. A similar principle applies in Boolean logic, except the operators are the logical operators such as AND and OR.

The functioning of those logical operators can then be illustrated by the use of a *truth table*. As an example, the truth table for the AND operator in Boolean logic is:

A	B	A AND B
0	0	0
0	1	0
1	0	0
1	1	1

In the previous truth table, the two inputs, A and B, are listed in the two left-hand columns. The output of the AND operation is listed in the right-hand column. As you can see, the result of this operation is only equal to 1 when **both** of the inputs, A **and** B, are equal to 1.

Similarly, the truth table for the OR operator in Boolean logic is:

A	B	A OR B
0	0	0
0	1	1
1	0	1
1	1	1

As you can see, the result of this operation is equal to 1 when **either** of the inputs, A **or** B, is equal to 1.

The simplest Boolean operation is the NOT operator. This simply acts to invert the input value, as shown in the following truth table:

A	NOT A
0	1
1	0

As you can see from the previous truth table, the result of a NOT operation will be 1 if the input is 0, and it will be 0 if the input is equal to 1.

The electrical circuits of Claude Shannon

Boolean logic, developed in the mid-1800s, might well have vanished as nothing more than a footnote in history had it not been resurrected in America in 1938 by the mathematician Claude Shannon. At the time, Shannon was a 21-year-old master's degree student at MIT, studying electrical engineering.

To understand Shannon's breakthrough, it must be realised that Shannon – as a student of electrical engineering – was well-acquainted with electromechanical *relays*. A relay is a type of switch. We all know how a switch works, just like the switch that turns the light on and off in your room. A switch is a mechanical device which you can move with your finger to open or close an electrical circuit. When a switch is *open*, there is no direct path for current to flow through the switch, but when the switch is *closed*, current can flow between the two terminals:

A relay operates in the same way as a mechanical switch, except that the mechanical movement to open and close the switch is caused by an electromagnet attracting a piece of metal. In this way, an electric current can cause the switch to be closed – instead of your finger doing the job.

The following diagram shows how an electromagnetic relay is a form of switch:

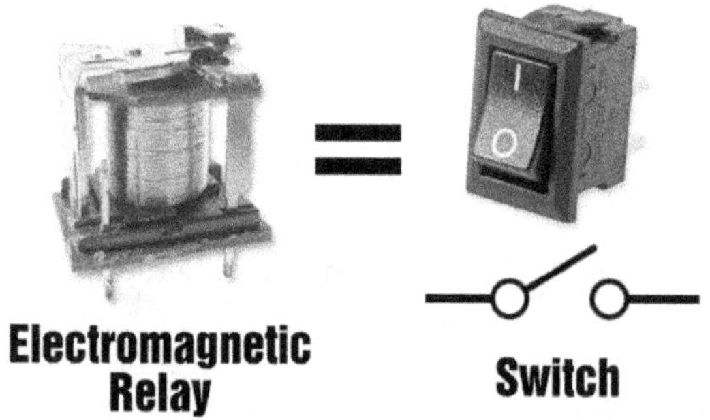

Electromagnetic Relay

Switch

Note that the mechanical switch on the right-hand side of the diagram has been labelled with the two values 0 and 1. This immediately reveals that a switch is a perfect device for processing Boolean logic. In the mechanical switch, the value 1 is being used to signify a closed circuit (current can flow, so the device would be turned on), and the value 0 is being used to signify an open circuit (no current can flow, so the device would be turned off).

Even though Claude Shannon was studying electrical engineering, he had studied Boolean logic in an undergraduate philosophy course, and this is where his breadth of knowledge proved invaluable (Shannon said later "It just happened that no one else was familiar with both fields at the same time"). Shannon realised that it would be possible to create Boolean logic operations using these relay switches. The idea was simple – brilliantly simple. As an example, Shannon realised that it would be possible to implement an AND operation simply by connecting two switches in *serial* ("serial" means connected in sequence, on the same wire):

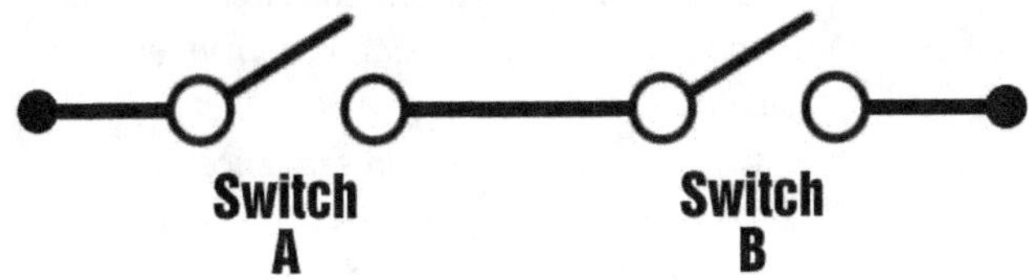

As you can see on the previous diagram, when two relay switches are connected in serial, then there will only be a connection along the total length of the wire when **both** switch A **and** switch B are closed. Therefore, this forms a simple implementation of a Boolean AND operation.

Similarly, two switches connected in *parallel* ("parallel" means they are connected side-by-side) would implement a Boolean OR operation:

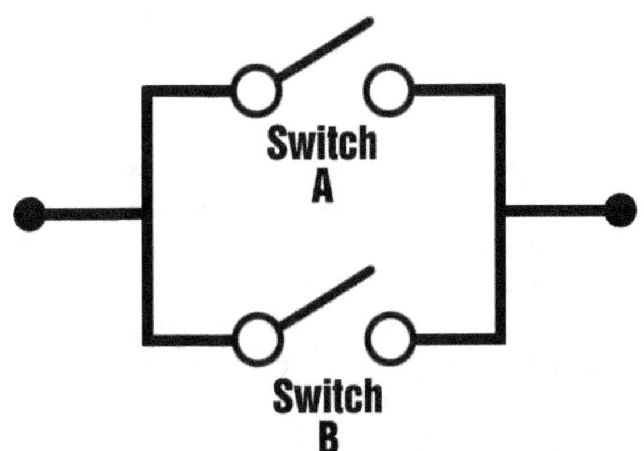

As you can see on the previous diagram, when two relay switches are connected in parallel, then there will be a connection along the total length of the wire when **either**

switch A **or** switch B is closed. Therefore, this forms a Boolean OR operation.

The biographer Walter Isaacson has written a book about the pioneers of computing titled *The Innovators*. In *The Innovators*, Isaacson describes how Shannon brought Boolean logic to electrical circuit design:

> *Shannon figured out that electrical circuits could execute these logical operations using an arrangement of on-off switches. To perform a AND function, for example, two switches could be put in sequence, so that both had to be on for electricity to flow. To perform an OR function, the switches could be in parallel so that electricity would flow if either of them was on. In other words, you could design a circuit containing a lot of relays that could perform, step-by-step, a sequence of logical tasks.*

Shannon had shown how electrical circuits could be designed to perform any logical function. According to the computer scientist Herman Goldstine, Shannon's work "helped to change digital circuit design from an art to a science". In his book *The Innovators*, Walter Isaacson describes Shannon's 1938 master's thesis as "the most influential master's thesis of all time". (Do not confuse this work of Shannon with his later groundbreaking work on information theory, which was described in my ninth book. The two different achievements are completely separate.)

As Walter Isaacson states at the end of his previous quote, Shannon had shown how to "design a circuit containing a lot of relays that could perform, step-by-step, a sequence of logical tasks". A device which can perform, step-by-step, a sequence of logical tasks is a computer. So Shannon had shown how a computer could be constructed using mechanical relays. Indeed, in 1941, the German engineer Konrad Zuse used 2,600 relays to build the Z3 computer. The Z3 was the world's first programmable digital computer (just to be clear, the Colossus computer which was made in Britain in 1944 – and was featured in my seventh book – was the world's first programmable **electronic** computer).

Here is a photograph of a replica of the Z3 computer on display at the Deutsches Museum in Munich. You can see the relays in the cabinet:

But the Z3 had a problem, a problem shared by all electromechanical computers. The root of the problem is that mechanical relays have moving parts, and moving parts are slow. As a result, the Z3 could only perform five to ten logical operations per second. We might compare this to a modern digital computer which can perform several billion instructions per second.

So where did the massive increase in speed come from? It came from a miniature device which was invented in 1947, as we shall now see …

Electronic logic gates

As has just been explained, mechanical relays could perform logical operations, but they were slow. The problem was that mechanical devices have moving parts. If a fast, practical computer was ever going to be developed, it needed a device which could perform logical operations without involving moving parts.

The solution arrived with the invention of the *transistor* in 1947. A transistor is a switch – just like a relay – but a transistor has no moving parts. Instead of the switching being achieved by the movement of a mechanical switch (by your finger, for example), a transistor is switched on and off by the application of a voltage to one of its three legs (the leg which is known as the *base*).

Here is a photo of a transistor (its actual size would be only about two centimetres long in total). Note the three legs:

The following diagram shows a transistor being switched on and off by the application of a voltage (effectively equivalent to a mechanical switch being switched on and off by a finger). The large black circle with the three black wires coming out of it is the standard representation of a transistor in circuit diagrams:

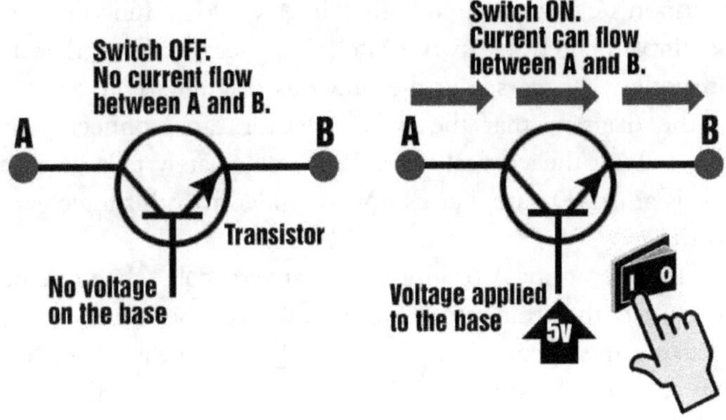

Just as Claude Shannon had shown it was possible to use mechanical switches to perform Boolean logical operations, so it was possible to use transistors as switches and arrange them into electronic circuits which could perform Boolean logic. Extremely fast computers could then be constructed.

These simple electronic circuits which can perform logical operations are called *logic gates*. The following diagram shows an actual circuit diagram which uses two transistors to implement an AND logic gate:

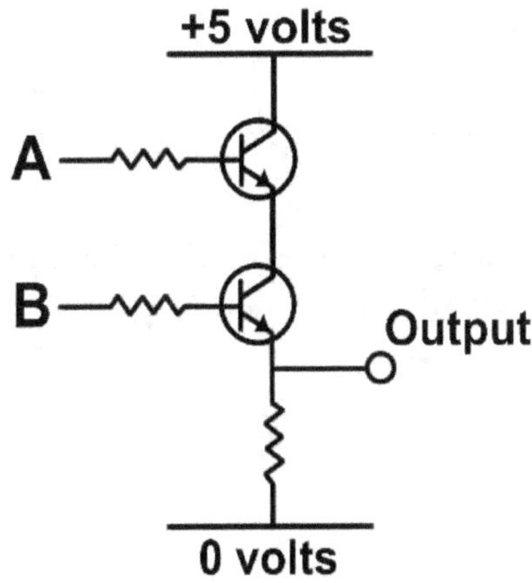

Remember from the previous section on Claude Shannon's electrical circuits that an AND logic function can be implemented by two switches in series. Well, that is precisely what we see in the previous diagram. You can see in the diagram that the two transistors are connected in series along the vertical wire. That immediately tells us that this is an AND gate. Let us now consider how this logic gate works.

If either input A or input B are at zero volts, then one or other of the transistors (switches) will be turned off, breaking the circuit along the vertical wire. In that case, you can see that the output will only be connected to the zero volt line (i.e., the output will be zero). This is the behaviour we would expect from an AND gate: if either one of the inputs is 0, then the output will be 0. However, if both of the inputs are 1, then both of the transistors (switches) will be turned on, and the output will be directly connected to the five volt line (i.e., the output will be 1). Again, this is the behaviour we would expect from an AND gate: if both of the inputs are 1, then the output should also be equal to 1.

In circuit diagrams, an AND gate is denoted by the following symbol:

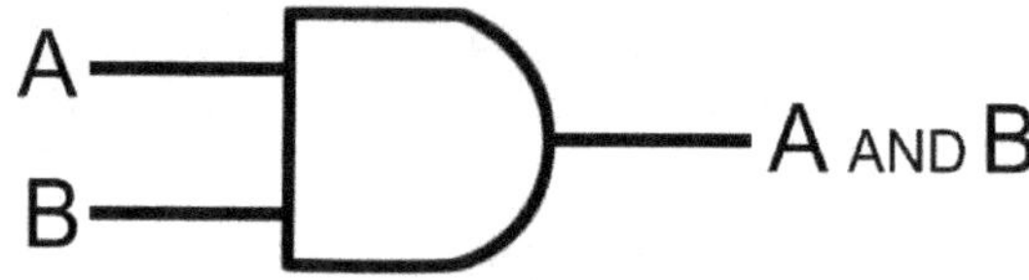

However, there is no reason why a logic gate should be restricted to having just two inputs. An AND gate with multiple inputs would look like this:

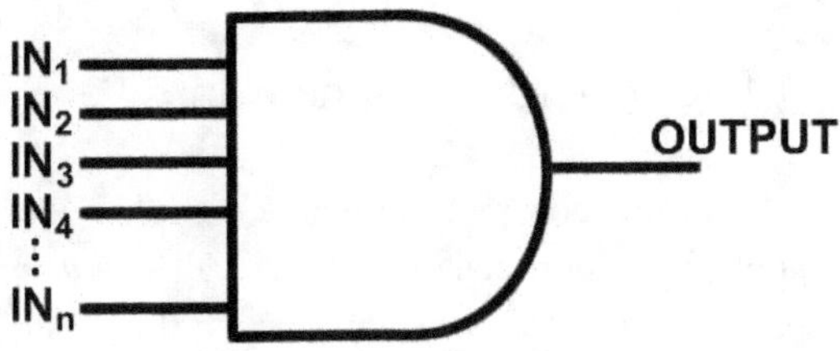

In the multiple input AND gate shown in the previous diagram, the output would be 1 only when **all** of the inputs are equal to 1, and would be 0 otherwise.

Note that – although there is no upper limit on the number of inputs to a logic gate – there is only ever one single output from a logic gate. The structure of a logic gate therefore mirrors the structure of an argument in logic. As described earlier, there is no upper limit on the number of premises in a logical argument, **but there is only ever one conclusion.** This principle is shown in the following argument which connects multiple premises using a multiple input AND operator. Note there is only a single conclusion:

IF
  I have a knife
AND
  I have a fork
AND
  I have a plate
AND
  I have some food
THEN
  I can eat the meal

And when we start to consider this structure of a multiple input logic gate, we find an uncanny similarity with the structure of another type of information processing unit …

The structure of a neuron

The brain is composed of an incredibly complex network of approximately 86 billion cells which are called *neurons*. In that sense, we might think of neurons as the "building blocks" of the brain. It is these neurons which transmit information throughout the brain, and allow information processing to take place. Somehow, it is the processing of information by these neurons which gives rise to consciousness.

The following diagram shows the structure of a neuron. As with all cells, a neuron has a cell body which contains a nucleus at its centre. Information enters the neuron via *dendrites*, which are branches which emerge from the cell body (shown on the following diagram). These dendrites receive input signals from previous neurons in the network. There may be in the region of a hundred dendrites on a single neuron. However, a neuron only ever has a single output, which is called an *axon*. Here is a simplified diagram of a neuron, only showing the features which are essential for this discussion:

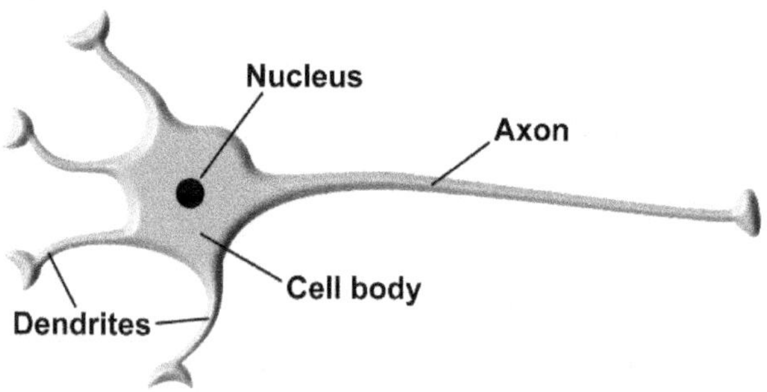

Inside the brain, the length of a neuron's axon might be as little as a millimetre. However, the axon which transmits information from a motor neuron in the spinal cord to the foot (telling a foot to move) can be a metre in length. This makes it clear that what we call "nerves" are, in fact, the axons of neurons (which are themselves "nerve cells"). To put it simply, all information in the body is transmitted by neurons.

Information is transmitted along an axon by a series of rapid spikes of electrical charge. These spikes can travel at a speed of over 100 metres per second. The mechanism which produces these spikes will be considered later in this book.

So, to sum up the structure of a neuron:

- Dendrites (inputs) bring information to the neuron, and there can be many dendrites on a single neuron.

- The axon (output) carries information out of the neuron, and **there is only ever one axon (output) on a neuron.**

The similarity with the logical structures we have considered so far in this chapter is clear. Truth tables, arguments in logic, and multi-input logic gates all have multiple inputs (no upper limit on their number) **but only one conclusion (output).** This is precisely the structure we see in neurons, with a large number of dendrites (inputs) but only ever one axon (output). It is as if logic is inscribed into the structure of each and every neuron.

But why should the structure of neurons necessarily take this logical form? It is because if you want to do anything useful with information processing, to solve some problem or puzzle, or to achieve some goal, or to come to some conclusion, then **you have to use logic.**

To illustrate this point, we might consider an example in which an early ancestor of *homo sapiens* might have been walking across the African savannah many millennia ago. Suddenly, he spots a lion on the horizon. He has to think of a way to escape from the hungry lion. In that case, he would have to use logic in order to devise a suitable escape strategy. His logical thought process might then be as follows:

IF
   I can reach the lake before the lion
AND
   I can swim faster than the lion
THEN
   I will escape from the lion.

So logic is required for any form of mental activity in which reasoning is required in order to do something useful, such as solve a puzzle, or build a house, or grow crops successfully, or escape a hungry lion. So, as an evolutionary necessity, **the human brain emerged as a device for processing logic.**

The reason for the logical structure of neurons then becomes clear.

Many great names in the history of science – including Alan Turing, George Boole, and Gottfried Leibniz – have believed that an understanding of logic is the key to understanding the working of the human mind. In the book *Introducing Logic*, we find the thoughts of the great German mathematician Gottfried Leibniz:

> *Logic is no longer a tool for convincing arguments but rather a system of rules of thought, so that even God's thought is necessarily logical. Even He could not create a world where a contradiction is true.*

In this chapter, we have seen how the human mind (and the mind of God) would necessarily have to follow the laws of logic. Presumably, even a demon would have to use the laws of logic. Might this, then, provide the clue we need to finally unlocking the mystery of Maxwell's Demon ...

8

LANDAUER'S PRINCIPLE

The green and leafy town of Yorktown Heights is located in upstate New York, about forty five miles from New York City. The peaceful nature of the town – coupled with its proximity to New York – has made it a desirable location for commuters, being just a fifty minute train ride from Grand Central Station.

Yorktown Heights is also where the research division of the giant computer corporation IBM has been located since 1961. Named after IBM's founder, the Thomas J. Watson Research Center has been responsible for some of the most important breakthroughs in computing, including the invention of dynamic random access memory (DRAM), the FORTRAN programming language, and the discovery of fractals (by Benoit Mandelbrot, while he was working for IBM at Yorktown Heights).

The period of the early 1960s was the start of the computer revolution. The research center at Yorktown Heights had been established with the aim of discovering the fundamental laws of computing.

One of the research scientists in the IBM facility was Rolf Landauer. As part of the programme to determine the limits of computing, Landauer was charged with exploring ways of reducing the amount of heat produced by a computer.

Here is a photograph of Rolf Landauer during his time at IBM:

Landauer successfully analysed the reason underlying the heat production of a computer. But his work also revealed a surprising side effect. It was to be that side effect which finally solved the problem of Maxwell's Demon.

And, as we shall now see, Landauer's solution had logic at its very core …

The irreversibility of logic

In order to analyse the production of heat by computers, Landauer considered how computers process information using logical operators. As we have seen in the previous chapter, electronic logic gates such as AND and OR form the core of information processing in a computer. Landauer noted something interesting about all of these logic gates. To understand what Landauer noticed, let us once again consider the truth table of an AND gate:

A	B	A AND B	
0	0	0	⎫
0	1	0	⎬ Same outputs
1	0	0	⎭
1	1	1	

Landauer realised that this truth table is irreversible. It is irreversible in the sense that once the logic gate has processed its inputs to calculate the output, you cannot then operate the logic gate in reverse to regenerate the inputs. We can see why this is the case from the previous truth table for the AND gate. If the output of the gate is 0, you can see that we cannot be certain if the input pairing had been 0 and 0, or 0 and 1, or 1 and 0 (on the previous truth table, you can see that those three conditions are grouped under the label "Same outputs"). So the logic gate cannot be operated in reverse as it is simply impossible to know what the inputs had been if you only know the output.

In his landmark 1961 paper, Landauer referred to this type of irreversibility as *logical irreversibility*. As Landauer said in his paper:

> *We shall call a device logically irreversible if the output of a device does not uniquely define the inputs. We believe that devices exhibiting logical irreversibility are essential to computing.*[1]

In this quote, Landauer is defining logical irreversibility as the situation in which "the output of a device does not uniquely define the inputs". This is precisely the situation we have been considering with the AND gate in which knowledge of the output does not allow you to determine the particular values of the inputs.

The second part of Landauer's quote is also interesting. By saying that "We believe that devices exhibiting logical irreversibility are essential to computing" he is essentially making the same point which was made in the previous chapter of this book. In that chapter, it was explained that if you want to do anything useful with information processing, then **you have to use logic.** In his belief that logical irreversibility is "essential to computing", Landauer is making the same point that logic is essential to computing.

In computing, you simply cannot avoid logic, and you therefore cannot avoid irreversibility.

---

[1] Rolf Landauer, *Irreversibility and Heat Generation in the Computing Process*, IBM Journal of Research and Development, 1961. **http://tinyurl.com/informationloss**

The inevitable production of heat

In his attempt to reduce the heat output of computers, Rolf Landauer realised that there would be a remarkable side effect of this logical irreversibility. To understand what Landauer realised, we need to return to consider Maxwell's Demon.

If you remember, the demon was able to reduce the entropy of the gas if he obtained **information** about the position and velocity of the atoms in the gas. Hence, it was explained that there is an inverse relationship between information and entropy. If information increases, then entropy decreases. And, vice versa, as stated in Chapter Six, "if information is lost, then entropy will increase".

So when Rolf Landauer examined these irreversible logic truth tables, he realised that information was, indeed, being lost in a logic gate. For example, a two-input AND gate takes two bits as input, and produces a single output bit. If you only have access to that output bit, you cannot then tell with certainty what the two inputs of the gate had been. This is due to the logical irreversibility of the gate. Before the operation of the logic gate, you knew the value of both of the inputs, but after the operation of the logic gate, you no longer know the value of both of the inputs. Hence, you do not know as much after the application of the gate as you did before the application of the gate. So some information would have been lost.

Another simple way of understanding that information must be lost in a logic gate is due to a logic gate having, for example, two input bits but only ever one output bit. Hence, one bit of information is lost. It is as simple as that. This point was made by Francis Crick in his book about consciousness titled *The Astonishing Hypothesis*. In the following quote from that book, Francis Crick compares the operation of neurons to logic gates, and realises that information must be lost:

> *One characteristic of a neuron is already fairly clear. A single neuron can fire at different rates and, to some extent, in different styles. Even so, in any period of time it can only send out limited information. Yet during that time the potential information coming into it, through its many synapses, is very large. In this process – going from its input to its output – **there must be a loss of information.***

As a result of this information loss, Landauer realised that entropy would increase (due to the inverse relationship between information and entropy). What form would this increase of entropy take? Well, it was explained in Chapter Five how Rudolf Clausius discovered the close connection between entropy and heat. Entropy is a measure of disorder, and disorder at the atomic scale is heat. So the increase of entropy would take the form of the production of an amount of heat.

How much heat would be produced? To answer that question we need to return to the result of Rudolf Clausius in Chapter Five. If you remember, Clausius defined the increase in the entropy of a system as:

$$\text{Entropy change} = \frac{\text{Heat}}{\text{Temperature}}$$

By reorganising this equation we get:

$$\text{Heat} = \text{Entropy change} \times \text{temperature}$$

So, to calculate the amount of heat produced, we simply need to multiply the increase of entropy by the ambient temperature. In Chapter Six it was explained that the amount of entropy associated with a single bit is $k\ln2$. So, for every single bit of information lost in a logic gate, Landauer realised that $kT\ln2$ joules of heat would be produced (where T is the ambient temperature).

This was a remarkable discovery, and it is called *Landauer's principle*. Computing pioneer Charles Bennett has said that Landauer's principle is "often regarded as the basic principle of the thermodynamics of information processing".[2]

I am convinced that the implications of Landauer's principle are not generally realised. Landauer had shown that any attempt to completely eliminate the heat produced by a computer would be doomed to failure – it is completely impossible. We might listen to the cooling fan spinning on our laptop computer and imagine that – given better technology or better design – the heat produced by the computer could be completely eliminated and so the fan would be unnecessary. However, that is not the case. Landauer had shown that whenever you have the logical processing of information, **heat must be produced.**

Rolf Landauer refers to this fact in his original 1961 paper, referring to the dissipation of heat in a computer:

> *The relevant point, however, is that the dissipation has a real function, and is not just an unnecessary nuisance.*

So heat must be produced by a computer. In fact, heat must be produced by any system which processes

---

[2] Charles Bennett, *Notes on Landauer's Principle, Reversible Computation, and Maxwell's Demon*, 2003.
http://arxiv.org/abs/physics/0210005

information in a logical manner, be it an iPad, or a digital watch – or a brain. In a nutshell, **logical information processing must produce heat.**

This production of heat is very noticeable in the human brain. The average power consumption of a human at rest is 100 watts. The brain has a power consumption of approximately 20 watts, which means that a fifth of the entire power consumption of a human is being used by the brain – and almost all of the energy used by the brain is converted into heat. Twenty watts represents a lot of heat (compare to a sixty watts incandescent light bulb – it is very hot).

In fact, if you were not aware of the true function of the brain, the production of heat would be its outstanding feature, and you might well imagine that was its main purpose. As an example, the philosopher Aristotle believed that the main function of the brain was for cooling the blood, with the brain acting as a radiator.

And this discussion about Landauer's principle and its implications (such as the production of heat) represents the final answer to the mystery of Maxwell's Demon. The entropy of the demon increases because of the processing of information in the demon's brain. The entropy increase would manifest itself as the production of heat from the brain of the demon. The entropy (disorder and randomness) of the heat produced by the brain of the demon would be more than enough to outweigh the reduction of entropy caused by the demon ordering the atoms of the gas.

The second episode of the excellent BBC documentary series *Order and Disorder* presented by Prof. Jim Al-Khalili considers Landauer's principle as a solution to Maxwell's Demon. Here is a link to a video on YouTube:

http://tinyurl.com/videomaxwell

The discussion of Maxwell's Demon and Landauer's principle is at the 52-minute mark.

The video considers the slightly different situation in which the information is not immediately lost in the logic gates. Instead, the demon stores that information in memory rather than deleting it. But, as the video explains, eventually the demon runs out of memory and has to delete the information after all – so the end result is the same.

Side effects

This discussion of Landauer's principle has revealed that there are four side effects associated with the logical processing of information.

Firstly, when we studied the truth tables of logical operators, we found that they were irreversible. Secondly, we have seen that this logical irreversibility leads to the loss of information across the gate. Thirdly, because of the inverse relationship between information and entropy, the loss of information leads to an increase of entropy. Finally, that increase of entropy would manifest itself as the production of heat from the gate.

The following diagram shows these four inevitable side effects of logical information processing:

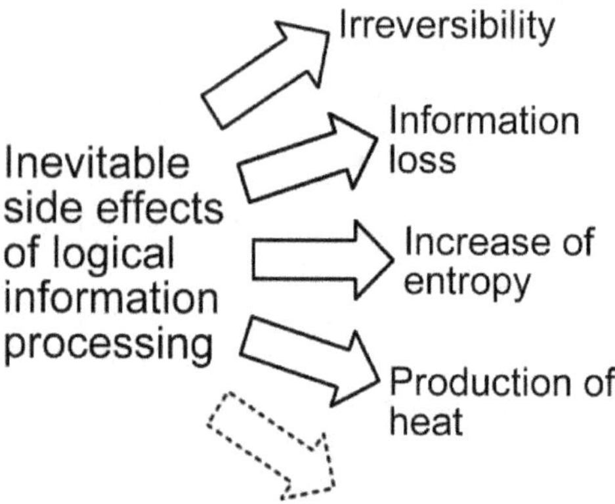

Note that a blank position has been included at the end of the diagram. Might there be a fifth side effect of logical information processing?

We will find out in the next chapter …

9

THE MAKING OF A THEORY OF CONCIOUSNESS

What do we mean by a "theory"?

Firstly, a scientific theory should be able to predict the outcome of experiments and observations with accuracy. As an example, one of the central theories in physics is the theory of general relativity, which is a theory of gravity. The theory can be used to predict the motions of the planets around the Sun with great accuracy. So a theory makes predictions.

This introduces another important property of a scientific theory: it can be falsified. For example, if I was to suggest a different theory of gravity, but my theory failed to correctly predict the motion of the planets (which general relativity predicts with great accuracy), then my theory would have been falsified. In that case, general relativity would have successfully repelled the challenge from my theory.

These aspects of a theory would apply to any potential theory of consciousness. The theory would need to make predictions about whether or not a system is conscious. That system might be a biological system (a brain) or an electronic or mechanical system.

But a theory should be more than just a generator of predictions. A theory should also provide an *explanation*. An explanation requires a deep understanding of **why** a particular phenomena occurs.

This point is made by the physicist David Deutsch in his book *The Fabric of Reality*. The first chapter of that book is devoted to describing why a theory is an explanation. According to Deutsch:

Our best theories embody deep explanations as well as accurate predictions.

I would also suggest that, to be an explanation which would satisfy a physicist, the explanation would ideally be phrased in terms of the behaviour of low-level, fundamental entities, such as particles, or fields, or space or time. A scientific explanation needs to be deep, referring to some low-level of reality.

Bearing all of this in mind, in the remaining chapters of this book, an original theory of consciousness will be presented. It will be based on the material we have covered so far in this book, in other words, it will have a firm basis in mainstream, orthodox physics. It will attempt to connect to reality at a deep level (the atomic level). It will be seen that the theory largely agrees with our intuitive notions about consciousness.

Perhaps, best of all, it is a simple theory.

OK, let's go …

An imaginary survey of physicists

I think it would be fascinating to organise a survey.

This survey would require the assistance of 100 of the leading theoretical physicists in the world today. Each physicist would be asked for his or her opinion as to what is the most likely solution to the mystery of consciousness. Theoretical physicists are a bright bunch of people, so you would think that they would know the answer, or at least the likeliest answer.

As far as I am aware, no such survey has ever been conducted, so we will have to use our imagination. Fortunately, I think I have a fair idea of the outcome.

Firstly, I suspect a large proportion of the physicists would refuse to answer the question, stating that they regard the subject of consciousness to be subjective and unscientific.

Secondly, I suspect a small percentage of the physicists would state that they agree with the controversial theory of consciousness presented by the physicist Sir Roger Penrose. Penrose's theory was first presented in his book *The Emperor's New Mind* in 1989. In his book, Penrose suggested that quantum mechanics plays the central role in generating consciousness (see my ninth book for details). Everybody loves Roger Penrose, but, unfortunately, I think it is fair to say that very few physicists agree with his idea. The small percentage who might agree with it in the survey would probably only be saying that because Sir Roger's theory is the only theory of consciousness that they know.

So, regarding my survey, up to this point it seems to be a case of "so far, so bad". The physicists do not seem to agree, and candidate theories – those theories which would be known by physicists – are few and far between, and suspected to be wrong.

But all is not lost.

Out of the physicists who provide an answer, who are prepared to suggest a likely solution, I am convinced one solution would have a huge lead in popularity over all the other proposed solutions. And it would simply be this:

Consciousness is an emergent phenomenon which arises from information processing.

You will notice the two key points in this sentence being the words "emergent" and "information processing". The word "emergent" simply refers to the idea that consciousness is a macroscopic effect – a human-scale effect arising from the interaction of trillions of microscopic particles. This was covered back in Chapter Four, so let us concentrate on the idea that consciousness arises from information processing.

Let us consider some of the high-profile physicists who would likely suggest this solution. In his book *Life 3.0*, Max Tegmark – a physicist at MIT – includes a chapter on consciousness. In that chapter, Tegmark states:

> *I'd been arguing for decades that consciousness is the way information feels when being processed in certain complex ways.*

So here is one example of a physicist suggesting that consciousness arises from information processing.

As another example, in a post on her popular blog *Backreaction* (in the post "Electrons Don't Think"), the outspoken physicist Sabine Hossenfelder is dismissive of any alternative explanation of consciousness:

> *Consciousness is an act of information processing. I don't understand why some people think there's something mysterious about it.*

So, as I suggested, a survey of 100 physicists would probably result in the majority view that consciousness arises from information processing. The reason why they should come to this conclusion is described by the physicist Paul Davies in his book *The Demon in the Machine*:

> *One thing isn't contentious: the brain processes information. It is therefore tempting to seek 'the source of consciousness' in the patterns of information swirling inside our heads.*

As Paul Davies says, the brain is an information processor, and our consciousness is certainly produced by the brain. It would therefore appear to be logical to imagine consciousness arising from the processing of information – maybe as a side effect.

In fact, that conclusion can seem to be so logical as to be almost beyond doubt. As James Stone says in his book *Principles of Neural Information Theory*:

> *Neurons communicate information,* **and that is pretty much all they do.**

On the basis of that quote, the proposed solution of the physicists can seem like the obvious conclusion:

1) There are 86 billion neurons in the brain.

2) All they do is process information.

3) Consciousness arises as a result.

4) Therefore, the logical conclusion is that consciousness arises from the processing of information in the brain.

Expressed in this way, it is clear why, I would suspect, this would be the most popular solution among physicists.

However, those physicists who suggest this solution would not necessarily know the details as to **how** consciousness arises from information processing. And that is the challenge for anyone with a physics background who is attempting to try to solve the mystery of consciousness. That is the challenge which we will attempt to solve in the remainder of this book.

And, as a first step towards that solution, we should take special notice of this idea that consciousness is a **side effect** of information processing. That would appear to be a crucial clue, as, in the previous chapter of this book, we have already encountered several side effects of information processing ...

The fifth side effect

I would suggest that there is a very important point which must be appreciated in the attempt to understand the link between information processing and consciousness. The point is that the information processing which is performed in the brain is not of some arbitrary nature. It must be, as described in Chapter Seven, **logical** information processing.

Once it is realised that the brain must perform logical information processing, then I think that is the first step towards a solution to this mystery. This is because – as explained in the previous chapter – any logical information processing system necessarily introduces Landauer's principle. The system would inevitably lose information in its logic gates (or neurons) and so it would be necessary to use Landauer's principle to fully describe the behaviour of the system. In that case, it might be said that the system *invokes* Landauer's principle.

And, as described in the previous chapter, if a system invokes Landauer's principle then side effects will inevitably be introduced. Those side effects would include irreversibility and heat production (a diagram of the four side effects was included in the previous chapter). And the idea of side effects appearing in an information processing system should make our ears prick up and set alarm bells ringing, because – as has just been explained – **many physicists suggest that consciousness is also a side effect of information processing.**

On that basis, it is a relatively small leap to suggest that consciousness is a fifth inevitable side effect of logical information processing, as shown in the following diagram:

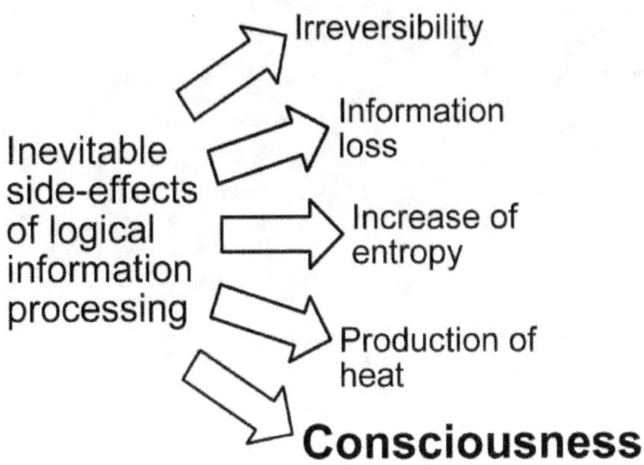

This idea forms the crux of the theory of consciousness described in this book. The idea is as follows:

Consciousness is an inevitable fifth side effect of logical information processing, as described by Landauer's principle.

Well, I told you it would be a simple theory!

Now, even though at the start of this chapter I stated that a theory should include an **explanation** of a phenomenon at a deep level, I am not going to include an explanation of the theory at this stage of this book. The proposed explanation will come later in this book. For now, this idea is being suggested as a tentative hypothesis which we will examine in the chapters to come.

In the following chapters, this idea will be used to test for the generation of consciousness in various systems.

10

DOES THE PRIMARY VISUAL CORTEX GENERATE CONSCIOUSNESS?

In 1952, the structure of the DNA molecule was discovered by Francis Crick and James Watson, who were working at the legendary Cavendish Laboratory in Cambridge, England. They were both awarded the Nobel Prize in 1962 for their discovery. However, it is not so well known that Francis Crick had an interest in solving another great mystery of biology. Crick was fascinated by the mystery of consciousness. And so it was that, in 1976, at the age of sixty, Francis Crick left Cambridge for California and started to pursue his interest in consciousness.

Crick started a collaboration with a young German-American neuroscientist named Christof Koch who was based at the California Institute of Technology (Caltech). It was to be a close and fruitful relationship which would continue until Crick's death in 2004. Together, Crick and Koch were determined to show the wider community that consciousness could be studied scientifically.

Crick and Koch wanted nothing to do with the endless philosophical discussions which had dogged the subject for centuries. Instead, they were only interested in developing **scientific** experiments which had the potential to shed light on consciousness. They chose to specialise on experiments which involved the human visual system.

One of the main reasons they chose to focus on vision was because it is easy to create optical illusions which provide insights into consciousness. In particular, it is possible to create optical illusions in which the consciously-perceived image can randomly switch between two different shapes.

As an example of this phenomenon, we might observe the following image of a vase, known as Rubin's vase:

As you stare at the vase, the image might suddenly switch so that you instead become consciously aware of the two faces staring at each other (the two faces on the side of the vase). Then, after a short period of time, the consciously-perceived image might switch so that you see the vase again.

But throughout all of this switching process, **the raw data presented to your eyes does not vary.** So the switching in your conscious perception must be entirely due to some switching in the activity of the neurons in your brain. Crick and Koch realised that if those switching neurons could be identified, then that would provide an insight into which parts of the brain were responsible for generating consciousness.

Francis Crick describes this approach in his 1995 book about consciousness titled *The Astonishing Hypothesis* (in the following quote, the "percept" means the perceived image):

> *Can we find neurons whose behaviour always correlates with the relevant visual percept? One way to do this would be to set up situations in which the visual information coming into the eyes remains the same but the percept changes. Which neurons change their firing, or style of firing, when the percept changes?*

It would appear that this approach could be used to identify the regions of the brain which are responsible for consciousness. Conversely, any region of the brain in which the neurons were **not** switching could then be deduced to be **not** playing a role in generating consciousness. And that resulted in Crick and Koch paying particular attention to the *primary visual cortex.*

The primary visual cortex is located at the back of the brain, and it is the first region of the brain to receive the raw visual data from the retina of the eye. The following diagram shows how the data is copied from the retina, along the optic nerve, to the primary visual cortex:

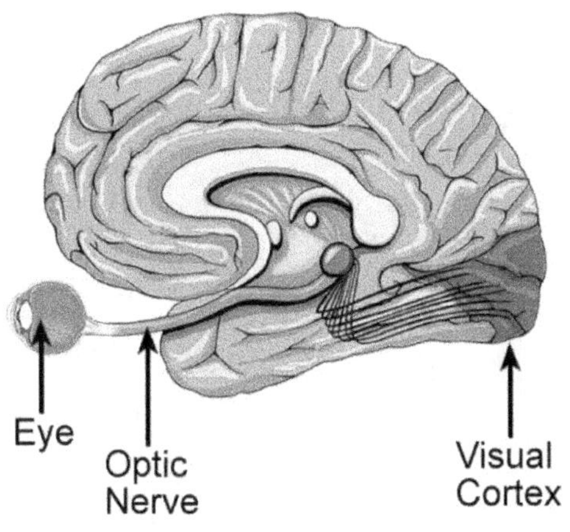

If we consider the Rubin's vase experiment again, we can see that the neurons in the retina contain the raw visual data from the image. Therefore, the activity of those neurons would not be switching as your conscious perception of the vase is switching.

The data in the primary visual cortex is just a copy of that data in the retinal neurons (copied along the optic nerve). Therefore, the activity of those neurons in the primary visual cortex would also not be switching as your conscious perception is switching. It would therefore appear logical to conclude that the primary visual cortex of the brain is not responsible for generating consciousness.

This conclusion of Crick and Koch is described by the physicist Max Tegmark in his book *Life 3.0*:

> *This same research controversially suggests that the primary visual cortex at the very back of the head is as unconscious as your eyeballs and your retinas.*

Therefore, according to the reasoning of Crick and Koch, the primary visual cortex is a region of the brain that does not generate consciousness. Let us now apply the simple theory of consciousness described in the previous chapter to see if it comes to the same conclusion.

A simple test for consciousness

Allow me to remind you of the theory of consciousness described in the previous chapter. The theory says that consciousness is an inevitable side effect of logical information processing – as long as that information processing invokes Landauer's principle.

So the central question we need to answer is: "Is Landauer's principle invoked by the information processing?" One way to answer that question is by examining the inputs and outputs of the system which are captured in the form of a truth table. The truth table is then examined to determine whether or not it is irreversible. As was described in Chapter Eight, irreversibility in a truth table is a sure sign of information loss – which leads to an increase of entropy, the production of heat, and the invocation of Landauer's principle.

The following rule is then applied:

- If the truth table is **reversible**, then the system is declared to be **NOT** consciousness-generating.

- If the truth table is **irreversible**, then the system **IS** declared to be consciousness-generating.

It is hard to imagine a simpler test for the generation of consciousness!

The whole of Chapter Three (and a song!) was devoted to this subject of irreversibility. Hopefully it is now becoming clearer how irreversibility could potentially play a central role in detecting the generation of consciousness.

Let us now apply this simple consciousness test to see if it predicts that the primary visual cortex is conscious or not conscious.

Applying the test

As we are initially stating that the retina and eyeball are not conscious (this is not controversial), the question we need to consider is whether the information processing which occurs when the information is copied from the retina to the primary visual cortex is the type of information processing which will generate consciousness. As has just been explained, we will answer this question by describing the inputs and outputs of the information processing operation in the form of a truth table.

A simple information copying operation applied to, say, one bit of information would have one bit of input and one bit of output. It turns out that there are two Boolean logic operations which have one bit of input and one bit of output, and they are the NOT gate and the Buffer gate. The NOT gate was considered in Chapter Seven. To remind you, the NOT operator simply acts to invert the input value, as shown in the following truth table:

A	NOT A
0	1
1	0

As you can see from the previous truth table, the result of a NOT operation will be 1 if the input is 0, and it will be 0 if the input is 1. Crucially, this truth table is **reversible**, meaning that if we know the value of the output then we know what the value of the input was. This is described in an entry in the online *Stanford Encyclopedia of Philosophy* by Owen Maroney (in an article which considers reversibility and irreversibility in logic gates):[3]

The NOT operation, for example, is a logically reversible operation. If the output is logical state one, then the input must have been logical state zero, and vice versa.

The other logic gate which has a single input and a single output is the Buffer gate. The Buffer gate simply takes the input value and copies it to the output, as described by the following truth table:

A	B
0	0
1	1

As you can see from the previous truth table, the result of the Buffer gate will be 0 if the input is 0, and 1 if the input is 1. So, as is the case with the NOT gate, this truth table is also reversible because if we know the value of the output then we immediately know what the value of the input was. So both the NOT gate and the Buffer gate are reversible.

The following diagram shows the symbols for both the NOT gate and the Buffer gate:

---

[3] Owen Maroney, *Information processing and thermodynamic entropy*, The Stanford Encyclopedia of Philosophy, **http://tinyurl.com/stanfordwebpage**

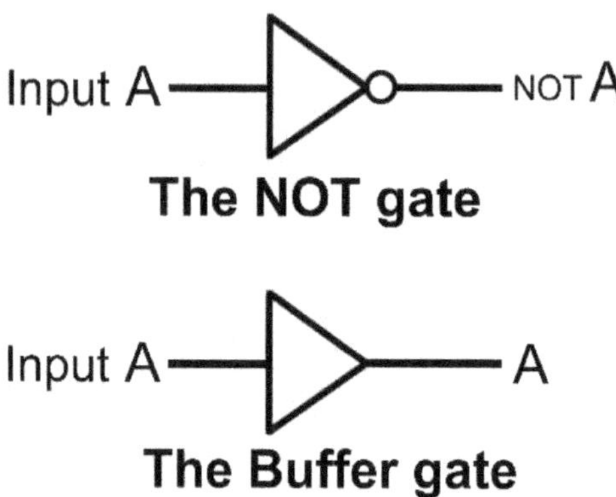

As you can see, the symbols are the same except that the Buffer gate does not have the small circle on the output of the gate (the small circle means that the output is inverted).

Let us now return to consider the primary visual cortex.

It has been explained that the information from the unconscious retina is copied to the primary visual cortex. A simple information copying operation would be represented by a Buffer logic gate. According to the consciousness test described earlier in this chapter, if the truth table describing a system is reversible, then the system is declared to be **not** consciousness-generating. As we have just seen that the truth table for the Buffer logic gate is reversible, we therefore declare that the information copying operation (from the retina to the primary visual cortex) has not generated consciousness. As a result, the primary visual cortex is declared to be unconscious – in agreement with the theory of Crick and Koch.

Copying information does not generate consciousness

As a general principle, this example has shown that the theory presented in this book predicts that simple copying of information from one point in space to another point in space (down a wire, for example) does not generate consciousness.[4] Clearly, as long as the information copying is error-free then the process is completely reversible: the output data could be sent back to the input, thereby perfectly restoring the initial input data. And, according to the theory, reversible operations do not generate consciousness.

Put simply, according to the theory, copying information does not generate consciousness.

----

[4] I found a *Physics Today* webpage which makes it clear that the copying of information does not invoke Landauer's Principle: "Gaining, or writing, information is akin to copying information from one place to another. Such one-to-one mapping can be realized, in principle, without dissipating any heat."
Information: From Maxwell's Demon to Landauer's Eraser,
http://tinyurl.com/copyinginformation

11

IS A THERMOSTAT CONSCIOUS?

I suspect the thermostat has been responsible for more arguments at home and at work than any other device ever invented. Certainly, that was my experience of office work. Just when everyone is feeling too hot, the thermostat decides to turn the heating on, and when everyone is feeling too cold, the thermostat decides to turn the heating off. It is as if a thermostat has a mind of its own.

And this idea – that a thermostat has a "mind of its own" – was given serious consideration by John McCarthy, a professor at Stanford University. McCarthy was one of the pioneers of artificial intelligence (AI). In 1959, McCarthy invented the LISP programming language for developing AI applications (the language is still widely used today). McCarthy was even responsible for coining the term "artificial intelligence" in 1956.

In 1979, McCarthy wrote a paper titled *Ascribing Mental Qualities to Machines*. In that paper, he wrote:

Machines as simple as thermostats can be said to have beliefs, and having beliefs seems to be a characteristic of most machines capable of problem-solving performance.

In his paper, McCarthy presented a real example from his house which had two thermostats: an upstairs thermostat and a downstairs thermostat. McCarthy imagined that his thermostats had "beliefs" as to the temperature of the house. In the example, it is clear that McCarthy had been having the usual thermostat-related problems:

*Recently, it was too hot upstairs, and the question arose as to whether the upstairs thermostat mistakenly **believed** it was too cold upstairs.*

The idea that a thermostat has beliefs and, therefore, consciousness has also been taken seriously by the leading consciousness philosopher David Chalmers. Chalmers considered the thermostat in Chapter Eight of his 1997 book *The Conscious Mind*. In that chapter, Chalmers followed a similar line of reasoning to McCarthy, suggesting that – because a thermostat can be considered as being an information-processing device (a device which can make decisions) – then it is not unreasonable to suggest it has consciousness. To justify his view, Chalmers considered moving down the animal kingdom, starting from humans:

To make the view seem less crazy, we can think about what might happen to experience as we move down the scale of complexity. We start with the familiar cases of humans, in which very complex information processing gives rise to our familiar complex experiences. Moving to less complex systems, there does not seem much reason to doubt that dogs are conscious, or even that mice are.

Moving down the scale through lizards and fish to slugs, similar considerations apply. As we move along the scale from fish and slugs through simple neural networks all the way to thermostats, where should consciousness wink out?

I believe the theory of consciousness presented in this book can provide an answer to David Chalmers' question. In fact, I believe a simple thermostat is a perfect example for revealing the point at which consciousness "winks out", as Chalmers describes it. In this chapter we will see how the application of the simple consciousness test can provide a clear-cut decision as to whether or not a thermostat is different from a slug.

So let us apply the simple test and see whether or not a thermostat possesses that vital property essential for consciousness.

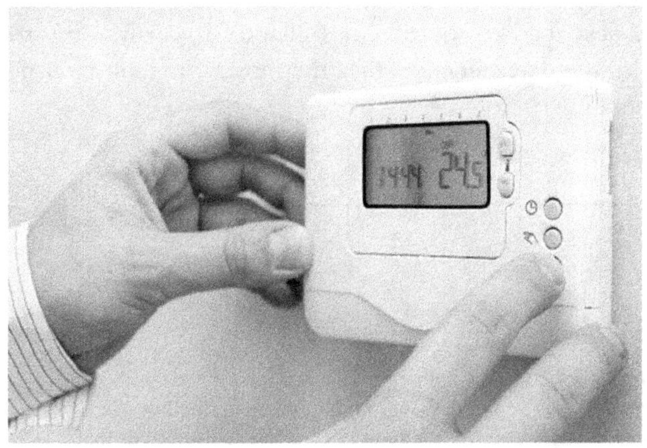

Applying the test

Let us now apply the simple truth-table based test for the generation of consciousness which was described in the previous chapter. Firstly, let us consider the behaviour of a thermostat.

A thermostat is a device capable of controlling the heating of a room. If the room is too cold, the thermostat turns the heating on, and if the room is too warm, the thermostat turns the heating off. A thermostat is therefore a device with a single input: the temperature of the room, and the thermostat has two possible output values: ON or OFF.[5] On the basis of the room temperature, the thermostat has to make a simple decision: "Should the heating be on, or should it be off?"

So let us construct the truth table which captures the behaviour of the thermostat. We can then apply the simple test described in the previous chapter to analyse the truth table to determine if the thermostat is consciousness-generating or not.

---

[5] In his book, David Chalmers considers a thermostat with three possible states corresponding to the room being too hot, the room being too cold, or the room temperature being OK. This implies that it is a more complex thermostat which is able to control the air conditioning to cool the room, as well as turn the heating on or off. This would require the setting of two temperatures to form an acceptable range. This would require two truth-tables: one to control the heating, and one to control the air-conditioning. But each individual truth-table would still be reversible.

For the purpose of capturing the functionality of the thermostat in a truth table, first let us consider the input values. Let us assign the value 0 with the room being cold, and the value 1 with the room being hot (we could assign these values the other way round as it would make no difference to the eventual conclusion – as will be explained in the final paragraph of this section). Then, considering the output of the truth table, let us assign the value 0 with the heating being OFF, and the value 1 with the heating being ON (again, these values could be the other way round). The truth table then becomes:

Room temp. (input)	Thermostat output
Cold 0	1 ON
Hot 1	0 OFF

We can see that what we have here is a logical NOT gate. When the input is 0 the output is 1, and when the input is 1 the output is 0. So the output is the inverse of the input. As explained in the previous chapter, a NOT operation is reversible because it is possible to "wind back" a NOT gate from the output to the input because if you know the output value then you instantly know the input value (it will be the inverse of the output value). So a NOT gate is reversible.

Finally, according to the simple test, a reversible truth-table implies that a thermostat is **not** consciousness-generating.

In other words, a thermostat is not conscious.

Note that if we had assigned the truth values the other way round when forming the truth table, then we would have been left with a Buffer gate rather than a NOT gate. As explained in the previous chapter, both of these types of logical operation are reversible, so the consciousness test would still have declared the thermostat to be unconscious.

A solid lump of metal

Perhaps this result can be best understood when it is realised that the job of a thermostat can be performed by a solid lump of metal. A thermostat can be constructed from two strips of different types of metal which have different expansion coefficients (a *bimetallic strip*). As the strip becomes warmer, the difference in the expansions of the two metals will result in the strip bending. If the strip is used in an electrical circuit, this can break the circuit and turn the heating off – as shown in the following diagram:

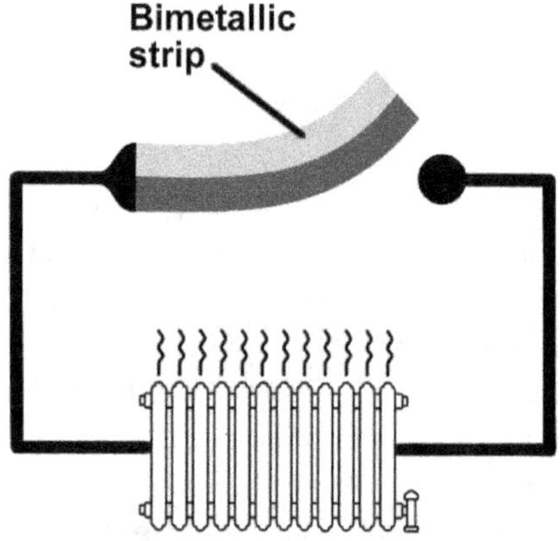

While we cannot be certain that a lump of solid metal is not conscious, this result certainly ties in with our intuition about consciousness: we don't generally believe that a lump of solid metal is conscious.

Another way of realising that this is a prediction of the theory is to appreciate that a lump of metal produces no internal heat. Heat production is one of the four side effects of Landauer's principle, and so, according to the theory, a system which is producing no heat from internal information processing is not consciousness-generating. This would eliminate all systems made of cold, solid material – such as statues and rocks – from being conscious. Again, this agrees with our intuition: we don't generally imagine that statues and rocks are conscious.

The test for consciousness described in this book therefore concludes that a thermostat – which might well be nothing more than a solid lump of metal – is not conscious.

The human thermostat

I would like to end this chapter with a rather technical – though important – point.

In this book, a simple test for the generation of consciousness has been presented. As we have seen, the test is based on the analysis of truth tables. However, we have to be careful of a potential pitfall when applying the test. Allow me to explain.

A truth table has several columns which represent the inputs to a system, and one column representing the output of a system. So when we analyse a truth table we are analysing the inputs and outputs of a system. In effect, we are treating the system like a "black box", with no regard for its internal functioning. Can it really be possible to detect consciousness purely on that basis?

It was the English mathematician Alan Turing who, in 1950, first considered the possibility of examining an information processing system purely on the basis of its inputs and outputs. He devised a famous test which we now call the *Turing test*.

In the Turing test, your task is to ask questions to two different information processing systems: a computer and a human. The computer and the human are hidden behind a screen so you do not know which responses are coming from the computer and which responses are coming from the human. The task of the computer is to imitate a human in its responses, and to fool you into thinking that it is a human.

At the end of the question-and-answer session, you are asked which is the human and which is the computer. So you have to make a decision purely on the basis of the answers you have received. If you mistakenly reply that the computer is actually the human, then the computer has achieved its goal of fooling you, and the computer is said to have passed the Turing test. According to Alan Turing, the computer could then be said to possess intelligence.

So, clearly, what we have here is an attempt to analyse an information processing system purely on the basis of the inputs and outputs of that system.

Now, we could surely imagine a highly-advanced computer which is capable of fooling you by producing identical responses to the human. In that case, the responses could have come from the computer, or they could have come from the human – we would have absolutely no way of distinguishing between the two different systems. This results in an important conclusion: **it is clearly not possible to uniquely define a complex information processing system purely on the basis of its inputs and outputs.**

Does this pose a serious problem for the simple consciousness test we have been using in this book? After all, that test is based on the analysis of truth tables, and truth tables just show the relationships between the inputs and outputs of a system. If the same truth table can describe two wildly different systems, then how can the method ever be used to reliably detect consciousness?

Fortunately, this is a problem we can overcome. The secret is that we have an advantage over a "black box" test such as the Turing test. Our advantage is that **we can open**

the black box and look inside. In other words, we have the ability to examine the internal information processing of the system – something we are prohibited from doing in the Turing test because we are not allowed to look behind the screen.

Our task then is to ensure that the truth table accurately represents the information processing which is being performed by the system. In particular, we have to be careful that the system is not a complex, conscious system which is just pretending to be a dumb, unconscious system.

As an extreme example of this, we might consider the hypothetical example of a "human thermostat". The human thermostat is a human who sits inside a big black box in your living room. The human has the simple job of turning the heating on and off when the room gets too hot or too cold (I would imagine job satisfaction would be rather low).

Purely on the basis of the inputs and outputs of that human-based system, we might construct the truth table for that system. The truth table would be the usual truth table for a thermostat which – as we have just seen – would be reversible. On that basis, the consciousness test would conclude that the system is not conscious, effectively deducing that the box contains nothing more than a lump of solid metal (a bimetallic strip) and is therefore not conscious. However, that would be an incorrect conclusion as the "human thermostat" is most certainly a conscious system!

So, as in the Turing test, we have to be careful when applying the consciousness test to check we are not being deceived. We have to "look inside the box" to ensure that the truth table is an accurate representation of the system.

That brings us to the end of this chapter.

In the last two chapters, we have applied the consciousness test to two very different information processing systems – the primary visual cortex, and a thermostat – and the test has returned a negative result in both cases. Before we consider the next example (which will provide us with our first positive result), there is something I must do.

At the start of Chapter Nine, it was explained how a scientific theory needs to do more than just make accurate predictions. The theory should also provide an *explanation*, it should explain **why** a particular phenomena occurs. Ideally, a scientific explanation should explain the phenomena in terms of low-level fundamental entities, such as particles, or fields, or space or time.

So, what is the explanation for the theory of consciousness which has been presented in this book? What is the connection between logical information processing and consciousness? Why is irreversibility so important?

I did not provide you with an explanation earlier, stating that an explanation would be provided later in the book. So I owe you an explanation, and – in the next two chapters – I will present an explanation.

Let me start by presenting what I feel is a very strong argument which provides a valuable insight into the nature and cause of consciousness …

12

THE SUSPENDED ANIMATION ARGUMENT

I am a huge fan of classic science fiction movies. I can remember going to the local cinema in the 1970s with my father – another sci-fi aficionado – to watch a stream of classic movies from *Planet of the Apes* to *Close Encounters of the Third Kind*. It was a golden age for intelligent sci-fi.

But there is always a problem with presenting science fiction in a movie in a way which is scientifically accurate, certainly a problem if you want to portray space travel. The problem arises simply because everywhere interesting in space is such a long distance away – and you cannot travel faster than light. If you have read my third book then you will appreciate that if there is one law of physics which we will almost certainly never be able to break it is our inability to travel faster than light.

As an example of the great distances involved in space travel, the star which is nearest to the Sun is Proxima Centauri. Proxima Centauri is orbited by two exoplanets which have the potential to support life (see my seventh book for details), so this would certainly be an interesting place to visit. Unfortunately, Proxima Centauri – the nearest interesting place outside our Solar System – is 4.2 light years away from us. So even if we could make a spaceship which travelled at the speed of light, it would still take a minimum of 4.2 years to get there. Considering more realistic speeds of travel, it would take longer than a human lifetime to get there.

So this raises a problem for any science fiction movie attempting to present a realistic portrayal of space travel. Put simply, it would take too long to get anywhere interesting in the galaxy. Film makers then have two choices. Firstly, they could ignore the speed of light limitation and employ fictional faster-than-light travel as in movies such as *Star Wars* ("hyperdrive") or *Star Trek* ("warp drive").

Or, alternatively, they could use the realistic solution …

Suspended animation

In February 2015, in the Pennsylvania town of Tresckow, 25-year-old Justin Smith went out for the night drinking with some friends. The temperature was twenty degrees Celsius below freezing, and there was snow on the ground.

At 9:30, Justin left the bar to walk home. At some point, it appears Justin blacked out and fell head first into a snowdrift. His body was not discovered until twelve hours later, frozen solid. There was no sign of a heartbeat. The paramedic on the scene reported back in a phone call and was recorded as saying "All signs lead us to believe that he has been dead for a considerable amount of time".

The state police arrived to investigate the death, and the body was transported to the local hospital. The team there were unable to pronounce Justin to be dead at that stage as the body was completely frozen.

However, almost working on a hunch, the hospital decided to try giving Justin a blood transfusion using warmed blood. After receiving the transfusion, early in the evening Justin's heart started to beat again.

Doctors were amazed to discover that Justin had suffered no brain damage, though some fingers and toes had to be amputated. To all intents and purposes, Justin Smith had come back from the dead.

The important factor in surviving being frozen appears to be the speed at which the body is frozen: not too fast (so that tissue is not damaged), and not too slow (so that the patient does not die from exposure before being frozen). If the speed of freezing is just right, then being frozen solid can actually save your life as it protects the body from the other harmful effects of exposure.

It is vital that the heart keeps beating – supplying the brain with oxygen – until the brain freezes. The brain is easily damaged by lack of oxygen, but if the brain is placed in suspended animation without being starved of oxygen at any stage of the process, then the brain will be undamaged. The patient can then be defrosted at a later stage without brain damage.

Fortunately, this is precisely what happened to Anna Bagenholm in Norway in 1999 when she was out skiing with friends. I can remember Anna's story making the international news reports at the time. Anna fell while skiing and plummeted head first through the ice covering a lake. Anna was trapped under the ice in the freezing water for a total of eighty minutes. Apparently her heart stopped beating after forty minutes, but that forty minutes in which the heart was supplying oxygen to the brain as it froze was crucial.

When Anna was eventually pulled out of the ice, she had no pulse, no breathing, and her pupils were dilated and unresponsive. She was clinically dead.

However, the Norwegians have a saying: "Nobody is dead until they are warm and dead". So Anna's blood was removed, warmed, and returned to her body. At 10pm that evening, Anna's heart started beating again. She has since made a full recovery.

These extraordinary stories provide a solution as to how to achieve long space journeys within the lifetime of an astronaut. The astronaut could be placed in a state of suspended animation at the start of the journey, and be revived many decades later when the destination is reached. This technique has been featured in many sci-fi movies including *Planet of the Apes*, *Alien*, and *Interstellar*. (Incidentally, *Interstellar* is the third Christopher Nolan movie to be mentioned in this book. Nolan certainly likes to get his science right.)

The following image is from the 1960s television series *Lost in Space* showing the Robinson Family being placed in suspended animation before a space flight:

NASA is seriously considering the possibility of using suspended animation for long space journeys. As part of its Innovative Advanced Concepts Program, research is being conducted to determine if astronauts could be successfully placed in a state of suspended animation for a journey to Mars.

Motion at the atomic scale

So, what are we actually doing when we place someone in a state of suspended animation? The real-life examples of Justin Smith and Anna Bagenholm which we have just considered have both involved extremely low temperatures. So what is it about low temperature that causes a subject to enter a state of suspended animation?

The physics behind heat and temperature was considered in Chapter Four. It was described how Rudolf Clausius, in 1857, was the first to realise that heat is nothing more than the motion of atoms in an object. If you rub your hands together, you cause the atoms in your hands to move, and your hands get warm as a result. Heat is nothing more than motion at the atomic scale.

So what we are actually doing when we freeze someone to place that person into suspended animation is to remove all motion from the atoms in their body – including their brain. So try not to think about hot and cold – this argument at its core is all about motion.

Now, here is an interesting question. When a person is in a state of suspended animation, with all of the atoms in their body motionless, do you think the person would be conscious? Surely not – it would be like the deepest sleep possible. So there appears to be a deep connection between motion at the atomic scale and consciousness: if motion is removed from atoms, consciousness is lost.

A similar argument was presented in 2017 by Romain Brette, a research director in theoretical and computational neuroscience at the Vision Institute in Paris.[6] Brette used the

1960s TV series *Bewitched* to illustrate the point. Brette considered a stoppage in the flow of time rather than a stoppage of the motion of atoms:

> *In the TV series Bewitched, Samantha the housewife twitches her nose and everyone freezes except her. Then she twitches her nose and everyone unfreezes, without noticing that anything happened. For them, time has effectively stopped. The question is: was anyone experiencing anything during that time? To me, it is clear that no one can experience anything if time is frozen.*

So, if motion is removed from atoms, it seems clear that consciousness is lost. The only possible counter-argument I can see would be the argument that a person whose atoms in their brain are stationary can still be conscious. In other words, a person with no brain activity can be conscious. Well, that's simply not the case, is it.

Let us perform a thought experiment to reveal the implication of this result.

Let us start by considering a conscious person. We are going to make a **single change** to this person, keeping all other variables unchanged. All we are going to do is use some advanced technology to render every atom in that person stationary (remember, this is a thought experiment so we can consider hypothetical situations even if we do not yet possess the suitable technology). No other change is made: the positions of the atoms are not altered, so no physical damage is caused to the person. Only the velocity of all the

---

[6] *Is a thermostat conscious?*
Romain Brette, **http://tinyurl.com/bewitchedbrain**

atoms which make up the person is stopped. At that point, according to the argument which has just been presented, the subject would no longer be conscious. So removal of motion at the atomic scale has resulted in a loss of consciousness.

We then make the second, single change. We restore the original motion to the atoms in the person. Again, no other change whatsoever is made – only atomic motion is added. As the state of the person has been restored to his or her original state, the person is now conscious again.

Logically, I can see only one conclusion. If you disagree with the following conclusion then you have to find a fault in the argument, and I can find no fault in the argument.

We have seen an unconscious, inert, unfeeling object become conscious when absolutely no other change is made apart from the addition of motion at the atomic scale. In other words, the addition of atomic motion has endowed inert, unfeeling matter with the gift of consciousness.

Therefore, it would appear that there can be only one logical conclusion:

Consciousness is caused by the motion of atoms.

I believe this is a hugely important result. I can see no logical weakness in this argument. I think it is watertight. Insights into the nature of consciousness which are fairly indisputable are few and far between.

It would appear that no arrangement of static, unmoving atoms can ever be used to create a conscious object. Consciousness is only generated when atomic motion is added to that object.

The result suggests that atomic motion is the essential ingredient for consciousness. It also suggests that if we want to discover the "stuff" of consciousness then we should look for a substance whose atoms are in continuous motion …

13

IS CONSCIOUSNESS A FORM OF HEAT?

As research for this series of books, I have been reading various books on the philosophy of consciousness. One common theme in those books is the suggestion by some philosophers that physics is unable to describe consciousness. After all, say the philosophers, there seems to be nothing in our knowledge of physics which resembles consciousness. As an example, David Chalmers devotes the third chapter of his book *The Conscious Mind* to "arguing that consciousness is not entailed by the physics of our world".

It would appear that these philosophers have a point. For example, if we look at the Standard Model of particle physics – our best model of fundamental reality – consciousness does not feature as a particle, or a property of any particle. As another example, we might imagine the construction of an enormous textbook containing our entire knowledge of physics. This hypothetical book might be called "The Big Book of Physics". The following image shows a copy of the Big Book:

If you then did a search for the word "consciousness" in the Big Book of Physics you would find nothing. So, on that basis, the philosophers would appear to be correct.

But what if there was another possibility. What if consciousness **was** included in the Big Book, but it simply did not appear in our word search.

Because it was called something else …

A proposed unification

In physics, progress in our knowledge has frequently been made by achieving *unifications*. A unification occurs when two phenomena – which previously were considered to be completely unrelated – are shown to be two different aspects of the same phenomena.

In Chapter Three, one the first unifications in physics was described. It was Newton who had shown that the force which holds the planets in their orbits is the same force which pulls an apple down from a tree. These two phenomena had previously been considered to be completely unrelated. Hence, by showing that they were two different applications of the force of gravity, Newton had achieved a unification. Another unification – briefly described in Chapter Six – was James Clerk Maxwell's unification of electricity and magnetism, showing that they were two aspects of the single electromagnetic force.

Unifications greatly advance our knowledge, while also simplifying our view of the universe. In a unification, two different theories are combined into a single, unified theory, which is usually simpler and clearer than the two theories which it replaced. Also, the unified theory describes reality at a deeper, more fundamental level than the previous two theories, and that can clarify our understanding of the universe.

Not all proposed unifications turn out to be correct. In his book *The Trouble With Physics*, Lee Smolin presented a history of unifications in physics. One of the proposed unifications involved the proposal for phlogiston in 1678, which was considered in Chapter Two of his book. As Lee Smolin describes:

Not all proposals for unification turn out to be true. At one time, chemists proposed that heat was a substance, like matter. It was called phlogiston. This concept unified heat and matter. But it was wrong. The right proposal for the unification of heat and matter is that heat is the energy in random motion of atoms.

Let us now return to consider the four inevitable side effects of logical information processing, as predicted by Landauer's principle (as was described in Chapter Eight). If you remember, those four side effects were irreversibility, information loss, increase of entropy, and the production of heat. Then, in Chapter Nine, a theory of consciousness was presented which proposed that consciousness was a fifth side effect of logical information processing.

So let us redraw the diagram showing the five side effects of information processing – but this time you will notice an addition to the diagram:

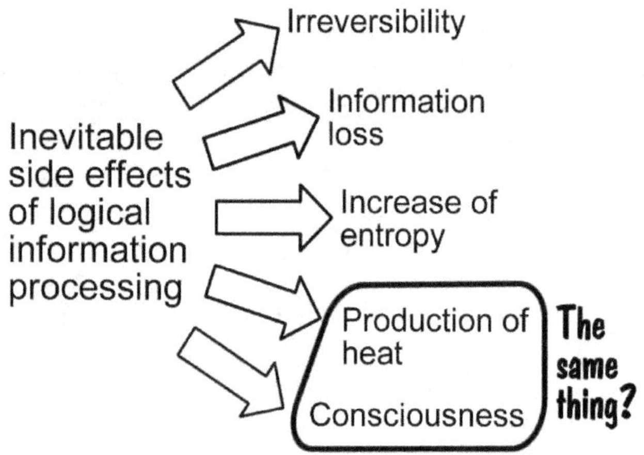

You will notice on the previous diagram that a loop has been drawn around two items: the "Production of heat" and "Consciousness". The reason for this is because I am suggesting a unification, a unification between heat and consciousness. Even though the production of heat appears to be a totally different phenomena from consciousness, I am suggesting that they are two different aspects of the same phenomena. In other words, they are the same thing.

This is a bold step which requires justification. Five reasons will now be presented why consciousness and heat are the same thing. I believe the last two reasons are particularly important.

Firstly, there are several qualitative similarities between our consciousness and heat. Our conscious experience feels complex, random, chaotic, ever-changing and unceasing – which is what you would expect if consciousness was the result of the complex, random, chaotic, ever-changing and unceasing interactions of trillions of atoms (which forms heat).

Secondly, as was explained at the end of Chapter Four, consciousness is a *macroscopic* property, meaning it is large, a part of the human experience, a part of the human world. Somehow, this macroscopic property emerges from the motion of *microscopic* particles in our brains. So, if we want to try to find consciousness in the "Big Book of Physics" – with the understanding that it might well be called something else – we should filter our search by only looking for macroscopic properties.

When we do that, we find that there are several macroscopic properties described in the Big Book. For example, if we have a sample of gas in a container, then that gas will have a certain pressure, temperature and volume. Each of those three properties is a macroscopic quantity. That is because each of those properties is a part of our human-scale world (we can measure the value of those quantities using our human-scale thermometers and pressure gauges). But each of those macroscopic properties would depend on the behaviour of the microscopic atoms which make up the sample of gas. So macroscopic quantities depend on the behaviour of their microscopic constituents.

These macroscopic properties are said to be "emergent" in that they only appear when a large number of particles are considered together. Individually, for example, it makes no sense to talk about the temperature or entropy of a single particle. Temperature and entropy are examples of properties which only appear when a vast number of particles are involved. The physicist Carlo Rovelli makes this point in his book *Reality Is Not What it Seems*. As Rovelli explains, 'hot' and 'cold' are macroscopic concepts which only emerge when a large number of particles are considered:

> *There are no 'hot' or 'cold' things at a microscopic level but, when we put together a large number of microscopic constituents and describe them in terms of averages, then the notion of 'heat' appears.*

So heat is a macroscopic property – just like consciousness. So at this early stage of our search, heat has passed our filter test as it is one of the macroscopic properties described in the Big Book of Physics. Heat has become a contender to be the "stuff" of our consciousness.

The third reason why consciousness might be a form of heat is because consciousness is an **irreversible** phenomenon. In Chapter Three it was explained how the fundamental laws of physics – and certainly the laws of Newton – are reversible in time. That means if we made a movie of the phenomenon, and then played that movie in reverse, the phenomenon would still appear to be correct. However, that is not the case for heat. Heat only ever flows in one direction, from hot to cold, so a movie of heat flow would not look correct if it was played in reverse. Similarly, our conscious experience has directionality in time: we feel as though we are moving from the past into the future, never vice versa. Also, we can remember the past but we cannot remember the future. If we somehow made a movie of our conscious experience, and played that movie in reverse, it would not appear to be correct (as the movie would show us remembering the future).

So heat and consciousness are both irreversible phenomena. This is surprising in a universe in which the fundamental laws of physics are reversible.

In a YouTube video, Carlo Rovelli has emphasized the irreversibility of consciousness:

http://tinyurl.com/rovellivideo

According to Carlo Rovelli:

*Consciousness is a process — it is something that happens. It definitely happens in time, and in a time-oriented way. It's an extremely **irreversible** process.*

Where does this time orientation come from? The ability to make decisions about the future, but not to make decisions about the past?

Why do we have memories about the past and not the future? The only answer is because of entropy.

Let us now consider the fourth reason why consciousness might be a form of heat. As described in Chapter Eight, Landauer's principle reveals that heat is an inevitable side effect of logical information processing. If it is true that heat is consciousness, then that would provide a wonderfully simple explanation as to why consciousness emerges in our brains. Consciousness would be an inevitable side effect of thinking, because heat is an inevitable side effect of thinking. It would provide an elegant solution to the question of why consciousness is produced by the information processing in our brains.

Now let us move on to the fifth reason why heat might be consciousness, and that reason is provided by the suspended animation argument presented in the previous chapter. As I described, I feel it is a very strong argument, possibly watertight. If you remember, the conclusion of the argument was simple and impactful: consciousness is caused by the movement of atoms. And what is the movement of atoms? Well, of course, as described in Chapter Four, heat is nothing more than the movement of atoms. For example, if you rub your hands together then you will make the atoms in your hands move, and your hands will get warm as a result. So the idea of consciousness being a form of heat ties in very nicely with the conclusion of the suspended animation argument.

So these last two reasons have provided two entirely separate arguments coming to the conclusion that consciousness is a form of heat. We have an argument based on Landauer's principle, and we have the suspended animation argument.

The following diagram shows the two arguments both pointing to the same conclusion:

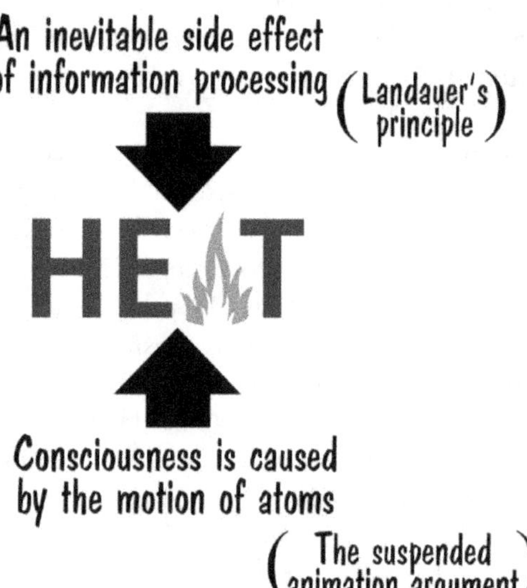

I would suggest that having two entirely independent paths leading to the same conclusion means that this theory of consciousness as a form of heat is a compelling theory. It suggests that the word "consciousness" **IS** included in the Big Book of Physics …

… but it is spelt "H–E–A–T".

Chaos and complexity at the nanoscale

To be clear, the type of "heat" being considered here is not truly heat at all – not in the generally-accepted meaning of the word. Heat, as defined in physics, is the truly random motion of atoms. The randomness is a crucial element. I am certainly not suggesting that truly random heat is a form of consciousness. No, the random heat we feel from the Sun or a campfire, for example, is surely not conscious.

However, the "heat" which is being considered in this chapter is the product of a very particular process: logical information processing. As such, the motion of the atoms would not be truly random. The particle motions would still retain correlations of the original information processing. Once the heat leaves the brain and interacts with the random motion of air particles, then those delicate correlations would be lost and at that stage it would truly become random disordered heat leaving our brains.

But while that heat is protected in our brains, the motion of the atoms would retain traces of the information processing which originally produced that heat. In other words, **it would retain traces of our thoughts.** It would be a special type of heat.

In Chapter Four, it was revealed how the atomic level is a extraordinarily fast-moving and violent environment. It was explained that one molecule of oxygen in a room full of oxygen gas will be moving on average at the speed of

461 metres per second, with molecular collisions occurring 6.8 billion times per second. All of this chaos is hidden from our eyes. It is easy to see how we might well overlook this churning, constant activity as being a potential cause of consciousness.

Perhaps we might speculate that microscopic transient structures might appear: order out of chaos. Maybe these structures might be formed by turbulence in the same way that tornadoes and hurricanes appear in the atmosphere. There is some interesting research in this area.

For example, in 2012, the physicist Peter Hoffman wrote a book titled *Life's Ratchet* which explores how biological systems exploit what Hoffman calls the "molecular storm" at the atomic scale. The use of the term "storm" seems to imply the possibility of the emergence of miniature "weather-related" structures such as microscopic tornadoes and hurricanes. Interestingly, Hoffman notes that the ancient Greek philosopher Democritus associated the human soul (consciousness?) with turbulence at the atomic level:

Democritus imagined the soul as a fire consisting of myriads of jostling particles, which Democritus called "atoms".

The similarity between Democritus' view and the theory described in this book is quite remarkable. Democritus' description of a "fire" as describing the soul is also interesting, making an early connection between consciousness and heat.

George Ellis, a professor of complex systems at Cape Town University, is also exploring the importance of molecular chaos in our thought processes. According to Ellis, the randomness of molecular chaos in neurons might lead to a way in which the apparent free will of our consciousness can be accommodated in physics. The concept of free will is notoriously difficult to explain using physics which describes the universe in a predictable, deterministic way. Maybe the chaos of swirling molecules provides an answer? George Ellis has described his theory in an online essay titled *From Chaos to Free Will.*[7]

Whatever is the case, according to the theory of consciousness described in this book, it appears that "messiness" at the atomic and molecular scale is required for consciousness: smoothness and order simply will not do the trick. This idea of the importance of atomic "messiness" will be explored in the example in the final chapter of this book.

But now, before we leave this chapter, it is interesting to note that this proposed close connection between consciousness and heat provides an instant, simple answer to one of the great scientific mysteries about consciousness …

Quantum mechanics and consciousness

When Erwin Schrödinger and Werner Heisenberg developed the theory of quantum mechanics in the mid-1920s, it launched one of the greatest revolutions in the history of physics. Applications of quantum mechanics include transistors, computers, and smartphones. It is the theory which has made our modern world.

However, when the theory was discovered, it quickly became clear that the role of the observer was particularly important. In 1927, the *Heisenberg uncertainty principle* revealed that the observer should not be regarded as being completely isolated from the system being observed. Instead, the act of observing inevitably affects the system. As an analogy from

____

[7] George Ellis, *From Chaos to Free Will*,
http://tinyurl.com/fromchaos

classical physics, when the air pressure in a tyre is measured, it is impossible to perform the measurement without extracting a small amount of air from the tyre. Therefore, the act of observation inevitably affects the system being observed.

In quantum mechanics, before a system is observed, that system can be in a remarkable state called a *superposition* state. For example, while it is in a superposition state, a particle can appear to be in two places at once (as revealed in the famous *double-slit experiment*).

It might sound bizarre, but this ability for a particle to be in two states at the same time is currently being used to make quantum computers. Quantum computers use particles in superposition states to perform multiple mathematical operations at the same time (for more details on quantum computing, see my tenth book).

So before a particle is observed, it can be in a superposition state. This superposition state can only be represented by a purely mathematical description called the *wavefunction*. However, the moment the system is observed, the system will no longer be in a superposition state. As an example, once it is observed, a particle will only ever be found at one particular point in space – not in two places at once. As a result, after observation there is said to be a *collapse of the wavefunction*.

This peculiar phenomenon forms the basis for the famous Schrödinger's Cat thought experiment. In the experiment, Schrödinger's cat is placed inside a sealed box. There is also a sealed glass jar of poison gas in the box, together with a small piece of radioactive material. If an atom of the radioactive material decays, the glass jar is broken, the poison is released, and the cat dies. Conversely, if an atom does not decay, the cat lives.

However, according to quantum mechanics, before the atom is observed we must consider the atom to be in a superposition state of both "decayed" and "not decayed". It appears we must therefore treat the cat as being both dead and alive at the same time!

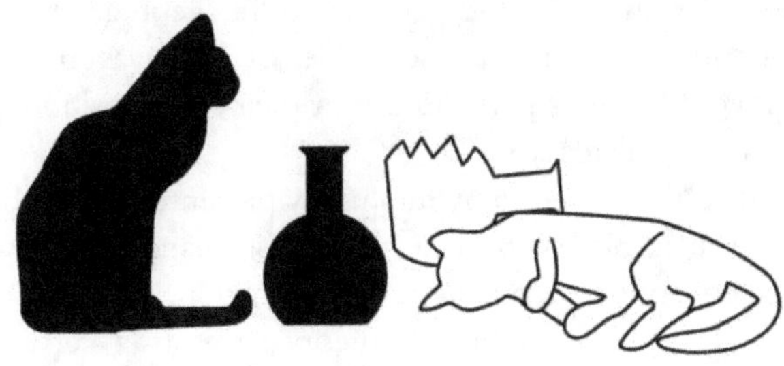

It is certainly true that if a human opens the box to observe the cat, the cat will be found to be either alive or dead – not in a strange mixture of both states. So there has been a "collapse of the wavefunction" caused by observation. The question was therefore raised, is there something special about observation by a conscious human being? Why do conscious beings never observe systems in superposition states? Why do conscious beings never observe cats which are both dead and alive? This has been one of the great scientific mysteries about consciousness.

Schrödinger's Cat remained a mystery for many years after it was first proposed. However, over the last fifty years, we have obtained a much greater understanding of the physical processes which occur during the so-called "collapse of the wavefunction". It is no longer so much of a mystery.

It appears that a particle which is in a superposition state is very rapidly reduced to a single, well-defined state when that particle interacts with a noisy, messy, random environment, full of fast-moving particles. This process is called *decoherence*. And what do we call the state when particles are moving in a random, fast-moving manner? That is heat. Heat is the random motion of particles. So a warm environment acts to collapse the wavefunction.

This is the reason why a quantum computer must operate at an incredibly low temperature. The temperature of the interior of a quantum computer is the lowest temperature in the entire universe, lower than the temperature of deep space. A quantum computer has to be kept at this low temperature to avoid the collapse of the wavefunction, which would take its particles out of their superposition state – ending the calculation.

And so this reveals why the theory presented in this book provides a simple solution of how consciousness will always act to collapse the quantum wavefunction. As has been explained in this chapter, the theory suggests a very close connection between consciousness and heat – even suggesting that heat is the "stuff" from which consciousness is made. That heat would therefore act to take any system out of its superposition state, very rapidly collapsing the wavefunction of that system.

In fact, according to the theory, it would be literally impossible for a conscious system to **not** collapse the wavefunction of a system it was observing. According to the theory, no conscious system could ever see a cat alive and dead at the same time.

The theory therefore provides a very simple solution as to how a conscious system will always collapse the quantum wavefunction.

That brings us to the end of this chapter. The next chapter is the final chapter of this book, and it contains another example of the simple consciousness test in action. It is also an important chapter.

Because the next chapter examines the big question …

14

DOES A NEURON GENERATE CONSCIOUSNESS?

The human brain is composed of approximately 86 billion neurons. Somehow, by connecting a large number of these relatively simple devices together, consciousness is created. On that basis, surely it would appear that the answer to the mystery of consciousness must be found in a neuron.

Why are neurons capable of generating consciousness? Why, for example, does Nature not make a brain out of 86 billion miniature solid metal cubes? What property does a neuron have that a solid metal cube does not have? If you can answer that question, I would suggest you would have solved the mystery of consciousness.

In this final chapter, an attempt will be made to answer that question.

We will discover some truly remarkable conclusions …

Applying the test

Let us start by applying the simple consciousness test to a neuron to see if it is predicted to be "consciousness-generating".

Of course, we already know what the outcome of this analysis should be. A neuron is certainly consciousness-generating. We know this because we know that if you have 86 billion neurons then you can make a brain (I am using this as the definition of what it means to be "consciousness-generating"). So let us apply the test and see if it comes to the correct conclusion.

Let me first remind you of the structure of a neuron.

A simplified diagram of a neuron was presented at the end of Chapter Seven, but it is presented again here. As you will remember, information enters the neuron via multiple *dendrites*. However, a neuron only ever has a single output, which is called an *axon*. Here is the simplified diagram of a neuron, showing the dendrites and the axon:

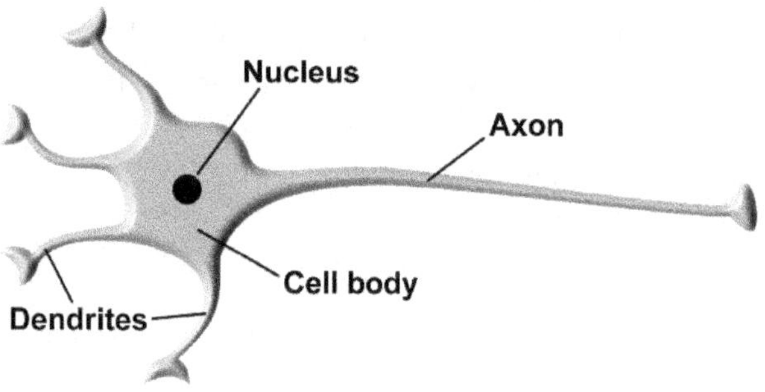

A neuron therefore operates like a logic gate, with the usual multiple inputs and single output of a logic gate (just as a truth table has multiple input columns and a single output column). This similarity between the functioning of neurons and logic gates has been noted by the physicist Paul Davies in his recent book *The Demon in the Machine*:

> *Because converging incoming signals from many neurons can be amalgamated, **the system acts rather like a logic circuit**, with the neuron being either on (firing) or off (quiescent), according to the state of combination of the incoming signals.*

As a neuron acts like a logic gate, we can create a truth table of its inputs and outputs and apply the simple test for consciousness-generation described in this book. We might, for example, consider a multiple-input neuron which is performing an AND logical operation:

A	B	A AND B	
0	0	0	}
0	1	0	Same outputs
1	0	0	
1	1	1	

As described in Chapter Eight, we can see from the previous truth table that this is an irreversible operation. As you can see from the truth table, if the output of the gate is 0, then we cannot be certain if the input pairing had been 0 and 0, or 0 and 1, or 1 and 0. So this logic gate could not be operated in reverse as it is impossible to know what the input values had been if you only know the output value.

The consciousness test predicts that an irreversible truth table is a sign of consciousness-generation. Therefore, the test correctly predicts that a neuron is consciousness-generating.

If the consciousness test had predicted a negative result – that a neuron cannot generate consciousness – then the theory could have been rejected as being wrong. That is the advantage of a scientific theory of consciousness which makes clear, testable predictions.

Information loss in neurons

The irreversibility of the information processing in a neuron inevitably leads to the loss of information. Another simple way of realising that information must be lost in a neuron is due to the fact that a neuron has multiple inputs but only a single output. This point was made by Francis Crick in a quotation which was presented in Chapter Eight:

> *One characteristic of a neuron is already fairly clear. A single neuron can fire at different rates and, to some extent, in different styles. Even so, in any period of time it can only send out limited information. Yet during that time the potential information coming into it, through its many synapses, is very large. In this process – going from its input to its output –* ***there must be a loss of information.***

Each "bit" of information lost in a neuron would produce an amount of heat equal to $kT\ln2$ joules of heat, according to Landauer's principle. The suggestion is that it is that heat which would be the "stuff" of consciousness.

You will note that Boltzmann's constant, k, is included in the value. At the end of Chapter Four, it was explained that Boltzmann's constant is "the bridge between the microscopic and macroscopic world". For this reason, it was suggested that any final theory of consciousness would likely include Boltzmann's constant as a means of explaining how the macroscopic experience of consciousness can emerge from the microscopic movement of particles. So the presence of Boltzmann's constant here is an encouraging sign for the theory of consciousness described in this book.

In the previous chapter it was suggested that this production of heat – this "molecular chaos" – is the essential ingredient of consciousness. In contrast, a system which is sterile and cold, and which lacks that "messiness" at the atomic level, can never be conscious.

So let us now leave neurons behind and consider a solid metal cube instead. When we consider a cold, sterile object – such as a solid metal cube – we find none of that essential molecular chaos. We have already considered a solid lump of metal back in Chapter Eleven when we considered the question of whether or not a thermostat is conscious. It was explained that the function of a thermostat can be performed by a strip of solid metal (a bimetallic strip). It was also explained how the resultant truth table for that thermostat would be reversible – and therefore predicted to be **not** consciousness-generating.

So lumps of solid metal do not produce consciousness – they are simply too cold and sterile, with none of the essential "messiness" at the atomic level. Whereas the information processing in a neuron is irreversible and therefore inevitably produces heat – that essential atomic "messiness". And this is how the theory provides an explanation as to why Nature can make a conscious system out of 86 billion neurons – but not out of 86 billion solid metal cubes.

The neuron and the steam engine

At this point, I could quite happily finish this chapter. The consciousness test applied to a neuron has made the correct prediction, and the associated explanation appears credible. However, when we start to dig deeper into the functioning of a neuron, we discover some truly remarkable conclusions.

So let us dig a bit deeper.

Let us start by considering the steps which result in the so-called "firing" of a neuron, in which a neuron transmits its electrical signal to the next neuron in the network.

Earlier in this chapter, it was explained that the input signals enter a neuron through the neuron's dendrites. The voltages from the various inputs are then summed inside the neuron. If the total voltage exceeds a certain threshold (the *threshold potential*), the neuron is said to *fire*. At that point, the neuron sends a burst of electrical voltage down its single output (axon).

After the neuron has fired, the voltage inside the neuron body is reset to its initial value (the *resting potential*). The process then starts again, with voltage building inside the neuron. As a result, firing is a rapid and repetitive process.

The following diagram shows this sequence of steps during the firing of a neuron:

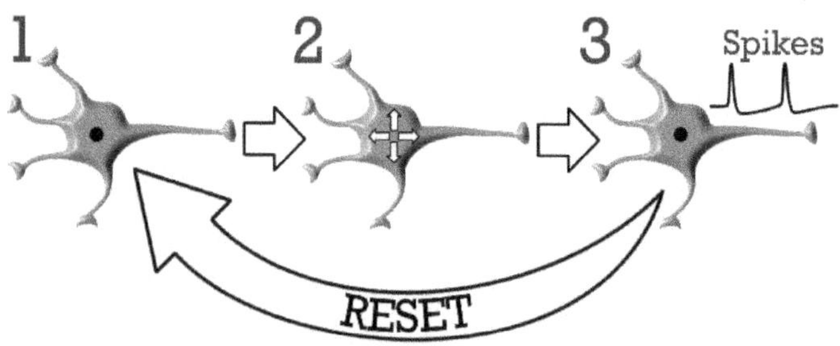

In the previous diagram, Step One shows the neuron in its initial state, with its internal voltage at its resting potential. Step Two shows the increase of internal voltage as electrical signals are received on the neuron's dendrites. Step Three shows the neuron firing, with a series of rapid voltage "spikes" being sent down the neuron's axon. Then the final stage is the reset operation which restores the neuron to its original state.

Because the sequence of steps ends in a reset operation, it is possible for the sequence to quickly repeat, and this is, indeed, what happens. A neuron will typically fire fifty times a second, with spikes of voltage being sent down the axon of a neuron like bullets from a machine gun.

So what we have here is a **cycle**. If you remember, we have already encountered cycles back at the start of Chapter One when we considered steam engines. A cycle was defined to be a sequence of linked steps ending with a reset operation. The reset operation is crucial as it allows the cycle to be rapidly repeated.

This definition of a cycle is described in the *Thermodynamics* textbook by Cengel and Boles:

> *A system is said to have undergone a **cycle** if it returns to its initial state at the end of the process. That is, for a cycle the initial and final states are identical.*

In particular, I would suggest that there is a remarkable similarity between the cycle of a neuron firing and the cycle which occurs during the jumping of the lid on young James Watt's cooking pot boiling on his mother's fire. To remind you, here again is the diagram of the jumping lid of James Watt's cooking pot:

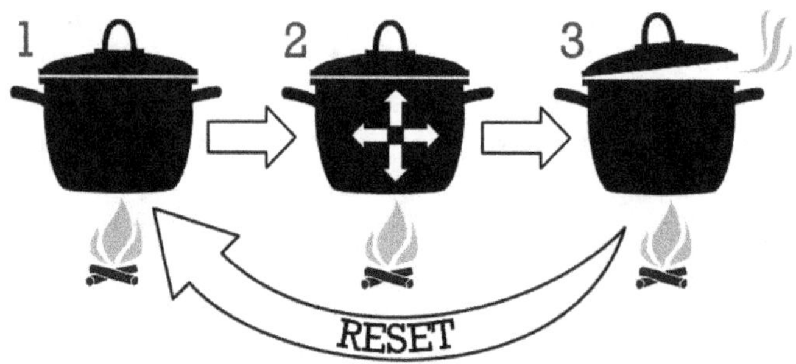

Initially, the pressure of the steam in the cooking pot is the same as the atmospheric pressure outside the pot. But as more heat energy enters the pot, the pressure of the steam in the pot increases (just as the voltage in a neuron increases). When the pressure of the steam in the pot exceeds a certain threshold value, it blows the lid of the pot up (just as the voltage inside a neuron has to exceed a certain threshold to initiate the spike). The whole process ends with a reset operation in which the pressure in the pot is equalised with the atmospheric pressure outside the pot (just as the voltage in a neuron is reset to its resting potential). Because of this reset operation, the lid of the pot drops back down again, and the cycle can rapidly repeat (just as a neuron firing can repeat fifty times a second).

In his recent book *Until the End of Time*, The physicist Brian Greene describes the importance of the reset operation in a steam engine:

> *The steam engine relies on a cyclical process: a piston is thrust forward by expanding steam and is then **reset** to its original position, where it awaits the next thrust. The steam, too, reverts to its original volume, temperature, and pressure, as must all of the engine's vital parts, readying the engine to heat back up and thrust the piston once again.*

So it appears that there are clear similarities between the firing of a neuron and the operation of a steam engine. And I would suggest that these similarities raise some interesting questions about neurons and consciousness.

Firstly, a steam engine does not produce a constant motive force throughout its entire cycle. Considering James Watt's cooking pot, we could only extract power from the movement of the lid at the point where the steam pushes the lid upwards. No force is generated as the lid drops down again. This is typical of all single-acting steam engines. This explains why, in a steam engine, it is essential that the cycle is rapidly repeated. As the thermodynamics textbook, *Concepts in Thermal Physics*, explains:

> *It has to be cyclic so that it can be continuously operated, producing a steady power.*

As a steam engine does not produce continuous power during its cycle, we might feel justified in asking a similar question about neurons: does a neuron generate consciousness continually? Or does a neuron only generate consciousness during a particular part of its cycle?

To answer that question, we must remember the reason why the consciousness test predicts that a neuron is consciousness-generating. The reason is because there is inevitable information loss in a neuron (not all of the multiple inputs can be passed to the single output). The point at which a neuron loses information is the point at which it fires. It is at that point that the voltage in the neuron is reset to its resting potential, with all internal information being erased. The spike of electric voltage down the axon would be matched by a burst of heat output (according to Landauer's Principle) and, according to the theory, a burst of consciousness.

Therefore, **it is proposed that a neuron only generates consciousness at the point at which it fires.**

Because consciousness is only generated at the point of firing, it would then be necessary for a neuron to fire repeatedly and rapidly in order to produce a continual experience of consciousness – just as a steam engine needs to cycle repeatedly and rapidly to produce continual motion. With neurons firing fifty times a second, the successive bursts of consciousness would generate a continuous impression of consciousness – just as frames in a movie played at fifty frames per second generate an impression of continuous motion.

The "reset" operation

There is another way to realise the potential importance of the firing process in generating consciousness.

When a neuron fires, its internal voltage is always reset to exactly the same original resting potential. This, therefore, acts as a "reset" operation. Rolf Landauer refers to a "reset" (or "restore") operation many times in his original 1961 paper, the paper which first introduced what we now call Landauer's principle.

In that paper, Landauer describes the reset operation as follows:

> *We start with each bit in one of two states and end up with a well-defined state.*

So, as Landauer explains, we have to create a truth table in which the inputs are in "one of two states" – 0 or 1 – but both inputs lead to the same single, "well-defined" output state. As an example, in the case of the following truth table for a reset function, both input values lead to a 0 output value (both inputs are reset to the same output value):

Input	Output
0	0
1	0

Applying the usual consciousness test, we now examine the truth table to see if it is reversible or irreversible. It can be seen that the truth table is irreversible (if we know the output is 0, we cannot be sure of the value of the input). The test predicts that an irreversible truth table is a sign of consciousness-generation. This is, therefore, another indicator that a neuron generates consciousness during its reset operation – the point at which it fires.

In his book about molecular biology, *Life's Ratchet*, the physicist Peter Hoffman considers several examples of "molecular machines". These incredible microscopic devices – only a few molecules in size – operate within the cells in our body. They are constantly in action, carrying nutrients around our cells, expelling waste, and repairing damage. But what is particularly interesting is that in Chapter Six of his book, it is explained how these molecular machines use cycles and reset operations to achieve their motion. In other words, this is conclusive evidence that the cells in our body – including our neurons – do, indeed, use the physics of James Watt's cooking pot.

Another example considered by Hoffman is the Maxwell's Demon thought experiment. Hoffman considers the situation in which the demon wants to process billions of particles – opening and shutting the trapdoor each time. As Hoffman explains, that represents a repeated operation: a cycle. In that case, there must be a reset operation in the demon's mind. As Hoffman explains:

In the case of the demon, erasing information restores or "resets" the system to its original state, allowing a new measurement cycle to begin.

The same applies to machines. To make a small machine perform repeated motions, a reset step is needed, as the machine needs to be returned to its original state before it can begin a new cycle. ***And it is this reset step that leads to an inevitable increase in entropy.***

Hoffman identifies the reset step as the point of increasing entropy – and therefore the point at which heat is produced. It is therefore the reset step that is potentially the point at which consciousness is generated in a neuron.

Neurons vs. electronic logic gates

Several times in this book, it has been stated that a neuron is a form of logic gate, performing the same function as an electronic logic gate in a computer. This is a well-accepted view, and it is largely correct. However, when we look more closely at the functioning of neurons and logic gates, we find there is one very big difference between them. And, based on the discussion in this chapter, it is a difference which might very well hold the key as to whether or not an electronic logic gate is capable of generating consciousness.

Let us first consider a neuron.

The firing mechanism of a neuron was considered earlier in this chapter. It was explained how the voltage in a neuron will keep increasing until the neuron fires – with the voltage inside the neuron being reset to the resting potential at that point. It is important to realise that this process in which the voltage continually builds will continue even if the inputs to the neuron are held steady and unchanging.

This important point is described in the textbook *Neuroscience: Exploring the Brain.*[8] In that book, a wonderful analogy to the process is presented. The firing mechanism of a neuron is compared to the motorised drive of a fashion photographer's expensive camera:

> *We likened the generation of an action potential to taking a photograph by pressing the shutter button on a camera. But what if the camera is one of those fancy motor-driven models that fashion and sports photographers use? In that case, continued pressure on the shutter button beyond threshold would cause the camera to shoot frame after frame. The same thing is true for a neuron. If, for example, we pass continuous current into a neuron through a microelectrode, we will generate not one, but many action potentials in succession.*

[8] *Neuroscience: Exploring the Brain*, Third Edition, 2007, Mark Bear, Barry Connors, and Michael Paradiso.

So, if the photographer keeps his or her finger depressed on the button, the camera will take a stream of rapid photographs. Similarly, if the inputs to a neuron are kept constant and unchanging, the neuron will still be an extremely active and dynamic object, continuing to send a stream of spikes down its axon at a rate of around fifty spikes per second (50 Hz).

In this sense, we see a neuron is more than just a logic gate. Instead, a neuron is a logic gate ... **with an additional rapid firing mechanism attached.**

However, when we consider an electronic logic gate, we discover a very different situation. A typical circuit diagram of an electronic logic gate was presented in Chapter Seven:

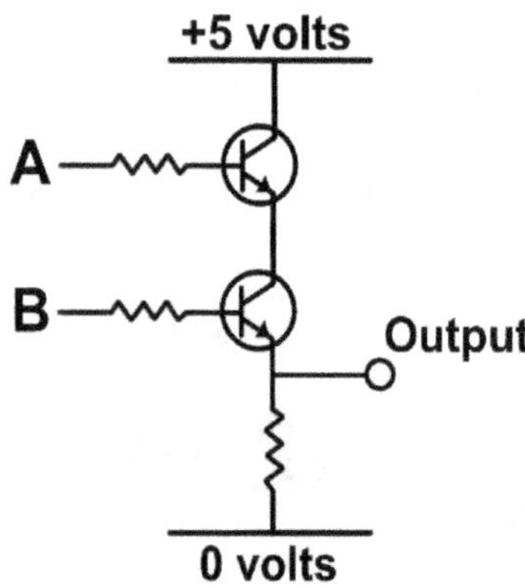

It can be seen from the diagram that this AND gate consists of just five electronic components: two transistors and three resistors. It is a very simple device, and this is typically the case for all electronic logic gates. Crucially, then, **an electronic logic gate does not possess the in-built rapid firing mechanism which a neuron possesses.**

As has been explained, if the inputs to a neuron are held at constant values, the neuron will continue its 50 Hz firing mechanism. In contrast, if the inputs to an electronic logic gate are held constant, then the output of the logic gate will be a continuous constant value.

This difference in behaviour between neurons and electronic logic gates is shown in the following diagram. In the neuron, it can be seen that the output will "pop" fifty times a second when the neuron fires: "pop, pop, pop, …" (I have used the terms "pop, pop" rather than "fire" to emphasize that this is a quick, repetitive action). Each of those "pops" will be accompanied by a loss of information, a burst of heat, and – according to the theory – a burst of consciousness. In contrast, the output of the electronic logic gate can be seen to be a constant value – with no "pops" of consciousness and heat:

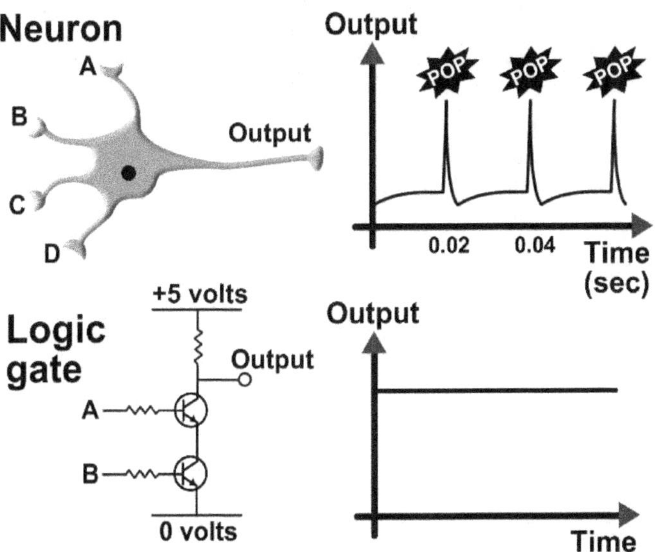

A neuron is simply a far more dynamic device than an electronic logic gate. As the suspended animation argument (in Chapter Twelve) suggests, movement is essential for consciousness. It would appear that dynamism is essential for consciousness.

Neurons are constantly dynamic, continuously losing information in rapid bursts. According to the theory, that results in a continuous experience of consciousness. In contrast, information is not lost in an electronic logic gate with static inputs: it would not generate consciousness.

In fact, it appears that neurons are the optimal consciousness-generating devices, almost as if they have been specifically selected for their ability to generate a continuous experience of consciousness. Why should that be the case? Is consciousness important? I have no answer to that question.

But I find that to be an absolutely fascinating conclusion.

PICTURE CREDITS

All photographs are public domain unless otherwise stated.

Photograph of William Murdoch's miniature steam locomotive is courtesy of the Birmingham Museums Trust and is provided by Wikimedia Commons.

Photograph of the Z3 computer is by Venusianer and is provided by Wikimedia Commons.

Press release image of *Lost in Space* is public domain, originally by CBS Television, and is provided by Wikimedia Commons.